YOU ARE NOT YOUR FARM

What the top 5% of farmers are doing that the average farmer is not

JACK OWEN

Copyright © 2024 by Jack Owen

All rights reserved. No part of this book may be reproduced or transmitted in any form or by any means, mechanical or electronic, including recording, photocopying, or by any retrieval system or information storage without prior written permission from the author.

ISBN# 978-0-646-89341-9

CONTENTS

Introduction	v
Chapter 1 Let's Define What Success is to You	1
Chapter 2 Principle #1: It is The Story That You Tell Yourself	9
Chapter 3 Principle #2: Vulnerability	23
Chapter 4 Principle #3: Health is Wealth	41
Chapter 5 Principle #4: The Power of Networking	55
Chapter 6 Principle #5: Gratitude	71
Chapter 7 Principle #6: Focus	85

Chapter 8   **107**
Principle #7: Learning

Chapter 9   **117**
Let's Talk Legacy

Chapter 10   **123**
Routines for Success

Chapter 11   **131**
Bringing it All Together

Acknowlegments   **135**

Resources   **137**

INTRODUCTION

Before we start, I want you to know that this isn't a book on how to grow great crops, have the fattest cattle or even the finest wool. There is plenty of information at your finger tips on how to achieve those outcomes. Don't get me wrong, those are all great achievements within a farming business, but are they really the characteristics of what makes you a successful farmer?

I know farmers who are doing amazing things with their land and their businesses but are still not satisfied with the results. In my view, from the outside looking in, they are very successful, but who cares what I think about them, what do they think about themselves? What is happening on the inside for them to be so unsatisfied? I do not claim to be a psychologist, far from it in fact, but I am intrigued by farmers. The way they think and act. How some have exponential growth, some are open to new ideas, and how some have done the same thing for four or more generations.

I love to question farmers, why do they do things this way and not that way, what drives them to keep farming through all the ups and downs, why this decision over that decision. I love hearing and understanding their journey they have been on. The obstacles they faced, how they overcome them, and what has made them the person they are today.

I find it fascinating that every farm is unique is its own way. From landscapes and layout, to enterprises and operation styles. Every farm is run slightly differently, even though as a collective, we are all producing similar products.

So how can it be that, two very similar farming businesses, with the exact same enterprises, with almost identical results, have two completely different reactions to those outcomes? I couldn't work it out and I really struggled to understand why. Having recently come back to the farm, pushing production was my main focus. However, no matter what result came my way, I always found a way to be dissatisfied with it. Why was this? It couldn't be from a lack of effort, could it? Do I need to work harder? Unfortunately, at the time, I thought this was the answer, to work harder. As I grew my network in agriculture, I was envious of farmers with similar results who were happy, relaxed, and had a sense of joy around what they were achieving. Why wasn't I like this? What am I not seeing? These questions made me frustrated, but through frustration came a break-through! Why don't I just ask them?

And now, here it is, I have put together this book based on what I have learnt from questioning successful farmers. The key principles you'll soon read about, are the exact principles that I discovered, that each successful farmer lives by. I now practice these principles as a result of the journey of discovery I have been on. This is my personal journey, but I truly believe this is what it takes to become a successful farmer. And the more I can share this with other like-minded farmers, I believe the future for farmers and generations of farmers to come looks amazing!

Why this book?

There is an abundance of self-help books out there, I know, I have brought almost all of them! Well, maybe not all of them, but I have close to fifty on my book shelf at the moment. From personal to business, the lot. I believed, that by reading them, success will just happen. I was a self-help addict. But nothing changed. I wasn't seeing the results I thought I was going to achieve. I couldn't relate to most of the content I was reading. Sure, there were some gold nuggets I found along the way, but I struggled to get the information to stick in my mind. Then I remembered an old high school teacher of mine. He told me that, if something isn't relevant to you, or if you cannot see the relevance in it, you will have a hard time comprehending it.

This teacher was my year 8 maths teacher. A subject I was struggling with at the time. He pulled me aside one day and said, I am willing to stay late every night after school and help you, if you actually want to do well in this subject. I had a lot of respect for this man, and I don't know if it was, I felt I needed to, or wanted to, but I agreed to stay back and give it a go.

The last bell of the day rang, and as everyone was leaving to go home, I was at my desk ready and waiting. He came in and the first thing he ask me was, 'what do you want to do when you leave school?'

In my mind I thought, was this maths tutoring? Does he even remember why I was here?

I replied, 'I want to be a farmer'

'And you enjoy farming?' he responded

'Yes! I love it in fact!'

'Okay, farming it is then'

He then proceeded to shift the maths equations, the same ones I couldn't understand just hours before to farming maths questions. I couldn't believe it! I started to understand these problems because it was connected to something I was interested in. A simple shift in thinking made it so much clearer for me. Likewise, a simple shift in relevance from solely self-help, to farming self-help, made understanding of what it takes to be successful so much clearer. I could now relate content to actual real-life situations. Situations that I have found to be common themes in the farming community, not only within Australia, but farming communities globally.

We have all made the choice to be a farmer. Whether that be through family and previous generations, marriage, or a start up with the passion of producing a high-quality product for others to enjoy. My goal is for you all to have an abundance of success in whatever it is that you're doing. This book is designed for you, the farmer, with real life relevance to what you may have gone through or may be going through. Everyone can achieve success; I have no doubt about that. So, sit back, relax, and let's get started!

So, what is success?

If we look at the oxford dictionary, it defines success as 'the accomplishment of an aim or purpose'. So as a farmer, what is your aim or purpose? And how do you define success? If you do not know what success looks like for you, then how do you know when you've achieved it?

The topics I talk about can be very overwhelming for farmers, almost put in the too hard basket. Or the old saying of 'I don't have time for this 'stuff', 'I have too many other 'jobs' to do'. But if you do not give this 'stuff' any attention, what will change in five, ten, or twenty years from now?

What you will read about can be quite difficult things to comprehend, but it isn't impossible! There is a very specific reason why you picked this book up to read, so if you do not understand something, go back and read it again, and again, as many times as necessary until it clicks for you. I guarantee you'll have a break through, just like me, and understand why this is so important, and how then to implement it into your own life. You'll be blown away by the impact of what some small positive changes can do for you. Now, I want to be clear, you must keep an open mind to what you're about to read. This book is designed to stretch you and your thinking, to challenge you and your thinking, and to create a new way of thinking!

Before we dive into the key principles of what it takes to become a successful farmer, it is important to have an understanding of what success means to you. On a farm business level but more importantly a personal level. Because if you do not know this, what is the point of doing what you do?

This may take some time to get a clear understanding of what success means to you. Be okay with-it changing overtime to adapt with your own personal growth.

I break it down into farm business success, and personal success, because everyone has a different idea of what success looks like. A farm business can be made up of multiple people who are all aiming for the same outcomes. However, those individuals will all have their own personal outcomes, that they would like to achieve in their own lives. I want to focus on personal success here, because it all starts with you. The most successful farm businesses are run by successful people. The success of the individual does not come from the success of their business, they worked on their personal success first, which has a flow on affect to their business. Read that last sentence again and let that sink in.

Success comes from within. You cannot buy success. You can definitely buy perceived success, but that does not make you successful. If someone has a successful farming business, as outsiders looking in, we automatically assume that they are successful. We give them that label. We may tell them they're successful and it may make them feel good. I am all for the praise of others and their achievements. I love seeing all people, especially farmers achieve. However, this is external success or material success. We are attaching success to a material thing, in this case, a farming business. How many of you have seen the situation of an older generation holding on to the farm until the day they die? They simply cannot let go. This is more common than you think! But why?

It is because they have attached their identity and success to that farm. They are the farm. Everyone in the community knows them and links them to that farm. They cannot let go because if they do, who are they? Not even they know the answer to that question because they either have never given it any thought, or do not want to think about it. Modern western society is elite at constructing the idea in all of us that material possessions equal success. Let me leave you with an idea to think about: 'you are not your farm'.

I couldn't tell you who owned our farm 150 years ago, and I cannot tell you who will own our farm in 150 years' time. Some of you may know the answer to the former in your own situation, but will struggle to answer the latter. You may like to think you know who will own it, but will that really have any relevance to you? Probably not is my guess. So, my question to you is, why do we identify ourselves to our farms?

In my opinion, it is what we know, and we are doing our best with the knowledge that we know. So, you cannot blame anyone, or look down upon anyone who attaches themselves to their material possessions like their farm. They are doing their best with what they believe is right. Just like I am doing my best with what I believe is right in creating this book for you. We are all on our own journeys through life, and it is not about what others are doing, but what 'you' are doing. What you are doing for yourself and the legacy that you will leave.

Breaking the stereotype of the farmer.

What comes to mind when you think of a farmer? For me, a hardworking individual with calloused hands who will work 7 days a week, and has adapted to accept anything that is thrown their way. But in society, we are known as complainers, aren't we? Too much rain, not enough rain, land prices too high, diesel prices too high, commodity prices are too low, farmers are never happy. 'The farmers will be right; the government will help them out yet again'.

It doesn't paint a very good picture for us does it. But does it have to be that way? I don't think so, actually I think quite the opposite. Did you notice that all the things I just wrote are all out of our control? We cannot control the weather, and I'm positive we cannot control what others think about us. So, what is stopping us from focusing on the things that we can control? Our attitudes, our language, the people we associate with, and our minds!

Now I can fully understand there are certain circumstances in farming that can have huge effects on our wellbeing. Droughts, floods and fires to name a few. They can come from nowhere and can definitely have a prolonged effect. However, we also have a choice on HOW we let them affect us, and HOW we choose to react to them. That is our choice! This may not be easy to hear, and I do not expect this to be easy, but sometimes we need to be challenged with new ideas, and it is healthy to challenge your thinking. Being open to new ways of thinking is the quickest and easiest way to grow.

Chapter 1

LET'S DEFINE WHAT SUCCESS IS TO YOU

How do I define my own success?

Remember – keep an open mind, and give yourself as much time as you need for this

I like to use Stephen R. Covey's approach from his book, 'The 7 Habits of Highly Effective People' and start with the end in mind.

Picture yourself looking back on your whole life, everything that has been and everything that could be. All the experiences you had, the memories you created, the people that were with you, what you achieved, the legacy you created, the memories that you feel would be most important to you. The memories that would bring you the greatest feeling of joy. What does it look like? What does it feel like? Remember, nothing is off limits in your imagination.

Now, write them out on a piece of paper. As many joy filled memories and experiences as you can possibly imagine.

It is never too late to begin this exercise. We all the ability to desire, and create our desires. If life isn't working out the way you imagined it to be right now, you feel trapped in specific situations and they are limiting your thoughts, start by imagining how you would like those specific situations to be instead. Play scenarios out in your mind with specific details of your desired outcome to these situations. Keep in mind, harming others in your own thoughts will only ever harm you. There is no room for revenge in the mind, if there is, only you will suffer. Allow everything you desire be filled with love and joy for yourself.

You now have a list of your perfect life, and it is truly what you imagined it to be. How do you feel reading it? Does it bring you a sense of joy? Does it put a smile on your face to know you could potentially achieve all these things? Yes? Then that's amazing! This is exactly how reading your list should make you feel.

If you are still struggling with this exercise, do not worry, please take your time with this. It may help you to get off your farm for a day or two and find a new, peaceful environment to be in. One that is away from your everyday distractions. Find a nice café to sit in, or head to the beach, and let your mind wander to truly get a sense of what it is you'd like in life.

With your list, lets now find common themes. Is there a common theme of family and friends? Or travel? How about the common theme of giving back?

Struggling to find common themes? If so, here's a tip for you. Find someone you trust. Get them to read from the start of this chapter so they have an understanding of what it is you're wanting to achieve. Hand them your list and allow them to find your common themes for you. Coming from someone with a different perspective is a great way to help you in this process so you do not get overwhelmed.

Once you find your common themes, these will now shape your values. Values in which you have chosen to live your life by, based on your desires of how you want your life to look. Keep the list of how you see your life handy, you will need this later on.

Example:

- Common theme of travel, you may value 'Freedom'
- Common theme of Family and Friends, you may value 'Love'
- Common theme of giving back, you may value 'Selflessness'

Now you have a list of your values, I want you to pin these up on your bathroom mirror so you will see them every day. This is your daily reminder that you are successful. However, it is your job to associate the action of these values with the feeling of success! This can be challenging at first, but practice makes perfect. Soon enough it will become second nature to you. Eventually you won't need to be reminded daily of your values. They will become you, in the way that you live each day. If the bathroom mirror is too 'out there' for you, make them your background wallpaper on your phone. I am sure you will see them quite often throughout the day.

But what does success feel like?

Think back to a time when you were a child and you were filled with pure joy and happiness. Nothing could have wiped the smile off your face in that moment. Winning a race at the school sports day, or having your named picked out of the hat for first prize in the raffle. That moment there! That is the feeling of success! Not winning the race itself, but the feeling as soon as you realised, you'd won! You felt on top of the world! Your feeling of success is unique to you and you only!

I remember back to when I was 11 years old. My parents had gifted me my first mob of sheep and it was time to sell their off spring for the year. My first ever sale! I was buzzing with excitement, I went to the saleyards and sat on the fence opposite them, waiting for the auctioneer to come around. Finally, my moment came, the auctioneer noticed they were mine and announced to the buyers who I was, pointed me out in the crowd and that this was my first ever sale. I couldn't wipe the smile from my face! The hammer dropped and I felt like a king! That was my feeling of success!

Every time you act on your values, there should be a moment of pure bliss. The secret is to feel it. Really embrace it, in fact to begin with, exaggerate the feeling! The more you practice this, you'll soon subconsciously be acting out your values without realising it, because the feeling of success will be all around you. You just have to trust the process.

If one of your values is freedom for example, and you want to act on that value, so you decide to book a holiday. This isn't a forced holiday of course, this is your choice, this has come from your values that you have decided upon remember. This has come from a place of the new version of you. In that exact moment of booking the holiday, feel that feeling of success start to flow through your body. When it comes from a place that is value driven, your specific values, it should put a smile on your face.

Now you have arrived at your holiday destination, take the time to go for a walk. Be alone and feel the freedom, feel being away from the farm.

Your mind may start to wander, and thoughts of how your farm is going, or 'did I shut that gate before I left' start to creep in. This is okay. It is okay for your mind to wander, especially when you're new to this. When this happens, thank the thought, because that's all it is. A thought. It is just a reminder to yourself to come back to the feeling of success. You hold the ability to choose a better thought to replace it with. 'I feel successful because I have acted upon my new value of freedom.' Say it with a smile on your face over and over. You may get strange looks from people whilst doing this but who cares! Why should you conform to a status of unhappiness or seriousness based on the perceived thoughts of strangers, people you don't even know, and are likely to never see again! This is for you! This is your holiday, from your vales, for your own life. You're the most important person in your life. Love or hate that last sentence as much as you like, but if you do not tend to yourself first, and give yourself self-love, you will be of no use to the ones closest to you. Why do you think the air hostess tells you to put your own oxygen mask on first, in the unlikely event of an emergency, before assisting others? For that exact reason! You will have no ability to assist others if you starve yourself of oxygen. You will become a liability, instead of being an asset.

I mentioned the words 'farm business success' earlier in this chapter. This is a whole new topic within itself. However, if you would like to know more about this subject, I highly recommend visiting Farm Owners Academy at farmownersacademy.com

Chapter Take-aways

1. To define what success looks like for you, you first need to define your values in which you choose to live your life by.

2. Use the exercise in this chapter to understand what your values are.

3. If you're struggling to find common themes, ask a trusted friend to read the chapter, and get them to help find common themes for you.

4. Once you have identified your values, put them up somewhere so you can see them every day.

5. Find what success feels like to you. Think back to early childhood memories that were filled with pure joy. Use the feeling of these memories, to associate the feeling of success, with your new defined values.

6. Every time you act through your new values, feel that feeling of success. Exaggerate the feeling if you must. This will help elevate your focus from external success to internal success.

Chapter 2

PRINCIPLE #1: IT IS THE STORY THAT YOU TELL YOURSELF

Mastering your thoughts

As farmers we are naturally busy people, or so we like to think we are. Make hay while the sun shines as the saying goes. But what is the reward at the end? Working one hundred hours a week in the busy times, even more for some! Then back down to forty or fifty hours a week in the quiet times. How many weekends have you had off in the past three months? I mean proper weekends off! Away from the farm, not one of those weekends where you 'check the sheep' or 'check the crops.' Don't get me wrong, there can be a lot of therapeutic benefits in doing those things, but only if you're one hundred percent present in that moment. That means your mind isn't racing around worrying about the gate that needs fixing when you drove past it, or the odd weed that is coming up amongst your crop. To be honest, if you're doing that in your time off, I'm sorry but you aren't actually taking time off.

Say you were to change careers altogether. You work from 7 am to 3:30pm, Monday to Friday. You get a twenty minute break for morning tea, half an hour break for lunch, and you're home by 4 pm each day. The typical standard 40 hour a week job. Sounds alright, doesn't it? Your boss has to ask your permission to work outside of those ordinary hours, and you're required to be compensated for any work outside of those ordinary hours. So why do you treat yourself differently? Sure, you may have more of a vested interest in your farm, you may own it, or your goal is to own it one day. Therefore, you may get the added benefits of capital appreciation, so that is why you go above and beyond in your role? But capital appreciation is only as good as what you do with it, right? People with regular 40 hour a week jobs have the ability to invest their money, and receive this benefit also, don't they? So, I am only focusing on the 'employee' side here. Which is what you are on your farm, if you work on it. So, why all of a sudden is there a different set of rules? Soon as you start working on your farm, are you not an employee? For Australian's, aren't you entitled to 4 weeks paid holiday per year? Aren't you entitled to a pay check each week? I want to really challenge your thinking on this one. Imagine writing a job advert for your role on your farm today. How would it read? 12 to 16 hours a day, Monday to Saturday, 8 hours a on a Sunday, with the bonus of Sunday afternoon off? How many applicants do you think would apply? Not many I would have thought. Yet, you agreed to it. You're agreeing to a standard that most people would run away from.

Now, I fully understand working hard to achieve an outcome or a goal. There are most definitely times in a farming calendar that require long days and big efforts from both you and your team. So, it wouldn't make much sense to pause your sowing program to take a holiday, then return to resume sowing. As a farmer, you understand there are very specific dates in which you must achieve tasks by, in order to set yourself up for a desired outcome. However, what is the goal you're wanting to achieve here? A goal has an end date attached to it. We will dive deeper into goals later on in this book, but I would like to bring this to your attention now. Yes, putting in the effort today to achieve a better tomorrow, or committing to a 5 or 10-year plan of solid work to set up a profitable business is amazing. But will you choose to enjoy it while you create it, and when you achieve it? Or will you be stuck in that groove for the rest of your life?

How about if I asked you what you were going to do today? Even if you didn't have anything planned, I bet you could rattle off 5 things off the top of your head that would satisfy my question. And I also bet they were all jobs to do with the farm. Why is that? Why do we feel the need to always be busy? Or have the persona that we are busy? Does busy equal success? Could it be the way we were raised from our parents? Or the way we would be perceived in the community if we weren't busy?

I think there are a lot of factors that play a part in this idea of always being busy. However, the underlying truth to all of this, is the story that we tell ourself.

There are eight billion people on this planet, and each of us have our own story that we tell ourselves. That is the uniqueness of the human being. Why do you think that one person can make one million dollars a year in passive income, while another person makes fifty thousand dollars a year working a forty hour a week job? It is the story that they tell themselves! Another word to describe this is 'paradigm'. It is the way they see the world. I like to think of a paradigm as a lens. Everyone has a different lens that they are using. This is exactly how the argument of I'm right, you're wrong is started. Because, we all think that the lens, the lens we have on, is the only lens. But what if I took my lens off and put yours on? Now I see the world the way you see the world. So, how do you see the world? This book for example, is about you putting on my lens, see what I see and see if you like it. If you like what you see, keep the lens on. So, what is the story that you tell yourself? Now, you may be reading this and saying, I don't have a story that I tell myself. Well, I assure you there is! You may not realise on a conscious level, but sub consciously, imbedded deep down there is an ingrained reason as to why you feel you need to be 'busy'.

From the day we were born, up to around the age of seven years old, is the most influential time for our subconscious mind to be programmed. We observe the world around us, and copy the behaviours of the people around us. Therefore, if we observed our parents working big days on the farm, no time off and always in go, go, go mode, then guess what! Like programming a hard drive in a computer, our subconscious has been programmed with the coding of 'work big days', 'no time off' and 'go, go, go'. Please do not be alarmed by this, because it is completely normal! It is no one's fault; it is just human nature. This isn't set in stone by the way; you aren't stuck with this for the rest of your life, and you have the ability to make the choice to change it. All you have to do is take your lens off and put a new one on. Remember our minds are similar to a computer, if you change the programming, you change the results.

A bi-product of the story that we tell ourselves, is our language we use. I don't mean the language of your country, but the language of your beliefs. Have you ever told someone something that you are going to do, or a goal or vision that you have? They reply with, 'you can't do that'. Then you start to think, 'well, maybe they're right, maybe I can't do that'. Do you know what just happened there? They answered with their beliefs! The lens they are looking through tells them that THEY cannot do it. So, if THEY cannot do it, they think it is impossible, therefore YOU cannot do it. Please be careful around these people. Be consciously aware of the language others use around you. Everything you hear is going in, and if you hear it often enough, you will start to believe it. You do not have to correct them, you simply have to say, in your mind, I do not believe what they are saying to be true.

If that is the language of others, what about your own language? What do you say to yourself, when you're by yourself? When I first started learning about the conversations, I have with myself, in my mind, I learnt that there are two words I must never use again.

'TRY' and 'CAN'T'

Unless you're a fan of the Rugby, you should remove the work 'try' out of your vocabulary immediately. It is a worthless word. If 'try' means to make an attempt to do something, then that means as long as we 'try', we will always succeed.

One of my goals in life is to have a positive impact on farmers worldwide. If I added the work 'try' in there. My goal would become; to 'try' to have a positive impact on farmers worldwide. My goal just became a thousand times easier to achieve, and I could easily achieve it in less than a day. I can think of a million ways to TRY and have a positive impact on farmers worldwide. I could attempt to write a book, like this one, and as long as I attempted to do it, then as the definition says, I would have achieved that outcome, right? My attempt could be as simple as opening my laptop, open a word document and write a title. Done. There's my attempt, can you say I didn't 'try'? Time to tick that life goal off the list and onto the next one.

If I was really passionate about farmers worldwide, do you think I would be satisfied with that outcome? No, definitely not. So that is why I no longer use the word 'try' in my vocabulary anymore.

Likewise, the word 'can't'. As soon as you say this word, your mind shuts off. You put an end to any possible thinking or problem solving. Replace the word 'can't, with 'how'. 'How' is an open-ended question. When something challenging comes up in your life, start by asking 'how'. This is being solutions focused. You will be surprised where your mind takes you when asking this. The possibilities are endless when you ask, 'how will I do this?' or 'how will I overcome this situation?' Instead of saying 'I can't do this'.

Remember to always be aware of what you're hearing and saying, words are very powerful things. Soon, your new open and positive words will just roll off your tongue with ease. I still catch myself today about to say 'try' and 'can't' on the odd occasion. All I do is smile and think back to the day I learnt why I do not use these words anymore. Then choose better words to replace them with. If you have fun with this new idea around language, it can be quite an easy thing to change in your life. all it requires is consistent awareness, practise, and time. Soon enough it will become common practise to you.

So how do I change my programme?

Now bear with me here, this will take an open mind to understand how to change your paradigm, but it isn't impossible! There are plenty of different methods used to change the way you think and see the world. So, I encourage you to explore as many ways as possible until you find something that works for you! However, the simplest way I have found to do this, that works for me, is the following:

1. Define the way you want to see the world. Be very clear and precise with this.
2. Become consciously aware of your thoughts and actions.
3. Question your thoughts and actions and understand where they are coming from.
4. Replace those thoughts and actions with your new paradigm or your new story you want to tell yourself.

Example:

1. I am so happy and grateful now that I have more than enough time to spend with my family and friends – defining your new paradigm.
2. I feel that after dinner tonight I need to go out and fix that piece of machinery that is broken down – Consciously aware of your thoughts and actions
3. Why do I feel I need to do this? That's right, growing up I often saw my parents having dinner then going back out to work to get a few extra jobs done before it got dark. Is this the way I want to live or am I like this because that is what I have seen to be true? – Questioning and understanding your thoughts and actions
4. If my new paradigm is to have more than enough time to spend with my family and friends, then I need to make the decision to spend time with my family and friends. So, I am going to consciously choose to spend time with my family tonight, rather than fixing that piece of machinery. – replacing your thoughts and actions with your new paradigm.

Let me guess,

'But if I don't fix that piece of machinery, we won't be able to get going first thing in the morning and we'll be even more behind!' Maybe that is true in certain circumstances, but I guarantee you it's not true more than 80% of the time. It is just the story you're telling yourself over and over again. Your mind will do anything to avoid change! So, it is up to you to become the master of your thoughts. This will take time and effort. You may want to start off small to begin with, then slowly increase over time. Allow for setbacks and embrace them. At the end of the day, you're only human, you're allowed to make mistakes. The biggest things with mistakes or setbacks are how you respond to them. Your old way of thinking will do anything to get you back into old habits, that is simply human psychology. My best advice is to feel the discomfort, then use the 5 second rule by *Mel Robbins*. That is, count down from 5 and make the decision that best serves you. The decision that you consciously know is the right one to make. 5, 4, 3, 2, 1 Go! Decision is made and immediately actioned. By counting down from 5 you give no time for your subconscious to kill it. Therefore, making a small step towards changing from your old paradigm to your new paradigm. Repetition and consistency are the keys to making this a success.

Let me put it to you in a different way. You come to the end of your life, what will you remember most, fixing that piece of machinery, working late every night, or seeing your partners face or your children's face's light up with joy the moment you make the decision to spend quality time with them, to create those unforgettable precious memories with them. Please have a long hard think about that. I am here to challenge your thinking! You may feel uncomfortable about it, but this is something you need to hear. Maybe this example does not fit into your life, and that is okay. How about calling a friend, or going for a walk? Use this time to read a book and learn something new. It is about doing something that fills you with joy. If it's baking a dessert and eating it with a bowl of ice-cream, then do that! I believe it is about ending the day with joy, closing your eyes to fall asleep with a smile on your face.

I think Henry David Thoreau said it best with:

"The price of anything is the amount of life you exchange for it" – Henry David Thoreau

We all have the ability to choose in life. That is the amazing thing about life! It is full of choices and we all must live with the results of our choices. Whether the results be positive or negative. So what choices are you making daily in your life right now, that results in positive outcomes? These should make you feel good! I urge you to keep doing those things. Keep building on those positive results. Likewise, what choices are you making daily that results in negative outcomes? I urge you to really re consider those choices. Are they serving you, or are you serving them? The best way to understand your negative choices that do not serve you is to write them out. Then to write the opposite of those choices next to them, and put a line through the negative choices as these no longer serve you. This adds strength to your new paradigm.

'~~I choose to not stop for a lunch break~~' – 'I choose to stop what I am doing and have 30 minutes for lunch every day to recharge'

'~~I choose to work every day so I can get more done~~' – 'I choose to give myself time to relax off farm, so I am at the top of my game whilst on farm'

'~~I choose to put the farm before my own health~~' – 'I choose to focus on my health first, allowing me to enjoy the farm for longer'

Every time you understand what you do not want, and construct it in a way to something that you do want, you're essentially re writing your program. You start to see things differently, you start to bring ease into your life and farm, you are starting to do things for YOU! This is the breakthrough right there. When you start to do things for you, as selfish as it may sound, you in fact lift the lives of the people around you. As a result of this you start to see your farm differently, rather than a chore, rather than a burden, it becomes a vehicle in which you can live your life the way YOU want to.

Let me challenge you again. Have you heard of the song 'Cats in the Cradle' by *Harry Chapin?* If you haven't, look it up, look up the meaning of this song. It is about a father who is too busy with his work and commitments to spend time with his son, this results in the son turning out just like his father, who in the end, didn't have time for. Now I am fortunate enough to have two parents who worked hard, but also gave every piece of their time and energy to their children. But how did you feel growing up? Did you become the second option to the farm? What messages are you sending to your children now? What messages would you like to send them? Or when you have the privilege of having children? Remember, from birth to around seven years old, children are like sponges. They observe and absorb everything!

Would you be happy if your children turned out just like you?

I have heard many farmers say, 'I am doing this so my children don't have to'. 'If I do this now, my children will be able to get a good start in farming or in life'. 'I want to allow my children to have more opportunities than I had'. I take my hat off to any farmer who says this, it is truly an amazing thing to do for your children, the future generation, and I strive every day to do the same for my children. But… there's always a but. What is more important in life? Your children seeing their parents happy, seeing their parents become grandparents and teaching their grand-children valuable skills around the farm, even enjoying special events like weddings and birthdays. Or seeing their parents working tirelessly their whole lives, so ingrained with the focus of 'we are doing this for our children' that they end up missing life's precious moments. I know what I would like my children to see. How about you? The question must be asked, 'what would you regret the most?'

You may be asking, well, can I have both? Can I help my children's future and also enjoy life's amazing gifts that come with having children? I have never seen a study saying, 'if you want a successful farm it will come at the cost of your family and friends' or 'if you want to spend time with your family and friends, your farm will go backwards'.

So, choose BOTH! Again, it comes down to the story you tell yourself! The choice is up to you. There is nothing wrong with saying, 'I have a successful farm and an abundance of time with my family and friends'. In fact, I want this for you! And you should want this for yourself too!

There is more to life than 'the farm'. The farm should be the vehicle that provides you with the ability to enjoy life. This can be an exciting new journey for a lot of you, so embrace the unknown, be consistent and remember to HAVE FUN with it.

Chapter Take-aways

1. Your old way of thinking may not be serving you, your loved ones or your farm.

2. Use the 4-step process to change your paradigm, if this doesn't feel right, do some research around paradigm shifts and find a way that works for you.

3. Use the 5 second rule when making a decision.

4. Write out your negative choices that do not serve you, cross them out and write the opposite.

5. What messages are you sending your children, and what messages would you like to send your children?

6. Create your new story and have fun with it!

Chapter 3

PRINCIPLE #2: VULNERABILITY

The lonely side to farming

This topic may be triggering to some readers, and I urge you to reach out for help if you feel the need to do so.

beyondblue.org.au
blackdoginstitute.org.au
mantlehealth.com.au

This is a tough topic to talk about, but I honestly believe it is not spoken about enough. Yes, in recent years it has become more common, and the more common this topic becomes the easier it will be to speak about it.

Farming can be quite an isolating environment to work in. Areas in north of Australia are hundreds of km's away from their closest town, you may not have great reception or the ability to be in contact with people for weeks, even months on end. However, you could be in close proximity to your neighbours, your country town, see people in the community daily and still feel isolated. Why is that? I had to know more about this. I had to find out why farmers in particular were having these feelings. What was going on for this to be so common. Being new to farming, was this something I had to evolve to become? Hide all emotion at all costs? Unfortunately, again, at the time, I thought this was the answer.

I am sure every farmer has a filing cabinet in their office, and if you were to open the top draw of that filing cabinet, and go right to the back of it, there'll be a folder. That folder will have no name attached to it, but it is essentially the 'hiding folder'. That's what I call it anyway. Any document you do not want to deal with, or think it's not that important now, but maybe one day I'll need that, you put it in this folder. Hiding it away, out of side, out of mind.

Well, that is what I created in my own body, a hiding folder. I would shove all my emotions down in there, as deep as I could, one on top of the other. It became quite easy actually. I was like a living robot. On the surface everything was like water off a duck's back. I wasn't processing anything; I wasn't feeling any emotions at all. Sure, I could laugh and smile, but when an emotion came through that I wasn't comfortable with, alarm bells rang! I do not have time to deal with this, I do not want to deal with this, see you later emotion, off to the filing cabinet you go.

One thing I didn't consider when building this emotional filing cabinet was what happens when it becomes full. How will I even know when it is full? Can it even get full? Aren't they just made-up things that go nowhere anyway?

Let me tell you, that filing cabinet got full alright, and when there is no more room in there, guess where it goes. It takes over and empties itself out. You have no control of it what so ever. This was one of the most challenging times in my life but one of the most amazing break-throughs to come from it. I was so shut off from people, I had lost the ability to express feelings, and I knew exactly why farmers do not like talking about emotions or struggle to express them. Why would I want to burden anyone with this 'stuff'. They have enough of their own 'stuff' going on, without adding mine to it.

I had an amazing insight come from this and it came from a chat with my business coach at the time, Tracy Secombe. Tracy is the author of 'From People Pleaser to Soul Pleaser', I highly recommend this book to anyone who can relate to my story.

So, Tracy and I got on a zoom call one day and I just couldn't hold it back anymore, I broke down, I couldn't speak. Finally, I got it together and asked, 'why am I the way I am?' I didn't even know who I was anymore. In her soft voice, in one simple sentence, she said *'you're being who you think you should be, not who you really are.'*

Let's back track a little here. I was new to farming, so I would always observe others. How did they do this, how did they do that. I would talk to veteran farmers; I would talk to as many as I could, to gather as much information out of them as possible. Little did I know, at the time, I was talking to farmers carrying large emotional filing cabinets. And as the saying goes, you're a product of the 5 people you hang around most. So, without my knowledge, it was inevitable that this would eventually happen, but a blessing in disguise none the less.

So, if I am being who I think I should be, then the burning question was 'who did I think I should be?' How do I find the answer to this? Honestly, I thought I was just me; this is who I am, what you see is what you get. Oh, how wrong I was. This led me to discover the power of vulnerability.

Vulnerability wasn't a part of my vocabulary. I actually didn't even know what the word meant. I had heard it used from time to time. A new born lamb is vulnerable to the elements or predators. The crop is vulnerable to a frost during its late stages of growth. However, I had never used or heard of the word when it came to myself.

Sitting at home one night, a message pops up on my phone, 'Men's health retreat – farmers only'. I didn't know what this was but for some reason I knew I had to be there; this was the answer to my burning question. The plus side was I knew all the other farmers going so I didn't think it would be so bad to have a catch up with them.

We arrived at an off-grid bush retreat site, just out of Hepburn in Victoria, Australia. The retreat was called 'Men's Breath & Break Through Retreat' run by Mark 'Tiger' Kluwer at markkluwer.com. Along with Trevor Hendy, an Australian ironman icon, and Kane Johnson, a former AFL superstar. I highly recommend checking out what these three individuals have to offer.

Fourteen farmers all camping in swags for three days. In other words, fourteen closed books camping in swags for three days. Or so I thought. This was a whole new experience for me, but an experience I had to have. I felt I learnt more in those three days about myself than I had in my previous years on Earth. Maybe a little exaggeration there but you get what I mean. This was an eye-opening experience for me and this set the foundations of writing this book. I finally had the breakthrough to my burning question.

The three days were filled with amazing conversations, powerful alone time, saunas, ice baths, cold plunges and breathing techniques. I could only have done this, if I had an open mind to it all. Similar to the open mind I request of you whilst reading this.

I want to speak about the key takeaways I took from this retreat, and how you, as farmers, can learn from them and how to implement them into your own lives.

Trevor Hendy ran a session on the topic he called, 'The Three Versions of Ourselves'. From the day you were born, right up to the breath you just took, every experience you have ever had, has formed the mould of who you are today. You are the product of your experiences.

The three versions of ourselves are as follows:

Our 'Front Cover'
This is a version of ourselves, of who we THINK people want us to be.

Our 'Back Cover'
This is everything that we are keeping inside.

And

Our 'True Self'
This is who we really are, which is caught between our front and back cover.

My front cover seemed to be as strong as the foundations on Hoover Dam. When I dove deep into my experiences, my front cover was made up of praise from achievements, the idea that everything is okay, I will do this on my own, I don't need help, high achiever, goal driven, disciplined, work hard, I'll rest when I'm dead, and I'll work even harder than before. These were all things I thought I had to bring to the table. If I was like this, then I would be accepted. If people saw this version of me, especially when I saw people seeing this version of me, it made me feel good! So, what's the problem? If it made me feel good, then isn't that a good thing? Well, I sure thought so, but look carefully. Not one thing in my front cover served me. No mention of my partner, my parents, my mates, the people close to me, the important things that truly matter in my life. I was neglecting importance to achieve status. My front cover was status. I was so caught up in what other people thought of me, that I lost sight of my true self. I do love my partner, my parents, my mates and the people close to me, but that all came second to my status in the farming community. Another word for this front cover is 'ego'

I have seen some huge egos in my life, hey, I just had to look in the mirror to find one of the biggest. In my opinion, egos in farming are the quickest was to ruin any farming business. Who wants to be around someone who is always right, can never see an alternative view, only listens to reply not to understand, and loves injecting themselves into every conversation? I'm sure no one walks away feeling enlightened when around these people.

I learnt a very valuable lesson when dealing with people with large egos. Because let's face it, you can run but you cannot hide. That's okay, they are who they are. You must learn to meet everyone with where they are at in their lives. So, every time I am talking to someone with a large ego who I cannot avoid, I do not agree or disagree with them, I simply use the line, 'that's interesting' and let them get out what they need to say. Have an understanding that this conversation is for them. The words they speak is for them. A large ego will speak because they have to say something, whereas the wise will speak because they have something to say. -paraphrased quote – Plato.

Post conversation, this is a time for you! Be aware of the words you just heard. I quickly reflect that none of that conversation served me, I am grateful for the conversation as a reminder to check my own ego, I give thanks that I am aware of my ego check, I let it go and move on.

On a side note, self-confidence and ego are two completely separate things. Ego stems from insecurities, hence our front covers which are full of insecurities. Where-as self-confidence is made up of awareness of self-strengths and accepting of self-weaknesses. Self-confidence is in our true self!

Onto my back cover now. Remember, this is everything that we are keeping inside ourselves. I didn't think I had a back cover anymore, hadn't I let it all out when my emotional filing cabinet decided to self-empty? Even though a lot had come up for me, there was still a back cover. To remove your back cover, you must process these emotions. Where were they coming from? Why did I not want to process them when they first come in? Do I have the ability to process them? I had a lot of questions around this as I seriously wanted to drop my back cover. Then Trevor dropped one of the biggest one liner's that I'm sure you may have heard, I'm certain I had heard it before, but hearing it this time, it really hit home for me.

<u>"It is okay to say you're not okay"</u>

Please read that last line again, and sit with it for a minute or two.

It is okay to say you're not okay. I think this is one of the hardest things to do in life but especially in farming, and especially if your front cover, who you THINK people want you to be, is still up as your protective shield. But why? Why is it so hard for farmers to say this? What is the missing piece of the puzzle that would make the words 'I am not okay' easier to say? Our front cover definitely doesn't help the situation, however, dropping the front and back cover work simultaneously. For us to drop the front cover, all we need to do is focus on dropping the back cover, and the front cover will inevitably be removed.

Through the busyness of being busy, I believe we have lost touch with connection. I remember getting home some nights, having had a tough day on the farm, maybe something broke or the sheep weren't making my life easy, and my partner could feel it. She would ask 'everything okay, you seem off'. SHEILD UP! My front cover would take over, you know, the one where everything in my life is perfect and that's how I think people want me to be? I'd reply, 'No, all good thank-you, nothing's up'. My front cover never wanted to bother others with burdens, but deep down, without me realising it, that was just another file added to that emotional filing cabinet I keep talking about. Plus, I didn't have time for that, I had too many thoughts in my head about what I was going to do on the farm tomorrow.

Mark 'Tiger' Kluwer then helped me understand how to drop my back cover. 'A problem shared, is a problem halved'. This is where vulnerability really comes in to play. Now, before I jump into this, let's have a chat about vulnerability. This is very important to understand. You get to choose who you be vulnerable with. Your vulnerability is not for everyone. This is your power and I want you to use it wisely. I had a very valuable lesson in this, a few days post retreat. I was vulnerable to someone who I thought I could trust, however, about two weeks later my story had come back to me via another person, who was using it as an ego driver for themselves. Lesson learnt! To have someone be vulnerable to you is a privilege and you must hold that privilege with the up most respect. It is for your ears and your ears only. I am sure you would like the same in return with your vulnerability. That is not to say you cannot help someone further after your conversation. Sometimes it is necessary to reach out for help if you are genuinely concerned about the individual. As they have built trust with you, by having this conversation, it is best to approach them first with your concerns. See if there is any additional help, they feel they need. You must understand that you cannot help someone who does not want to be helped. If you feel this is above your qualifications, show your vulnerability, express that to them, along with expressing your support in helping them to find the right support they may need.

How many of you have that select few people in your lives, that you could pick up the phone right now, call them, and say how are you going? Not for the typical answer of 'good, you?' I mean really asking, 'how are YOU going?' and they respond, 'well actually, I really struggled at work today.' If you have these types of people in your lives already, you're very fortunate, I really mean that. Now, there won't be ten of these people or even five. You may only have one, and one is all you need. One trusted people you can lean on.

Personally, I have three. I have three people that there is no BS with when it comes to these conversations. They can ask me anything, I can ask them anything, and I know that there are no front covers involved. It can be quite confronting at the start. Ringing someone and asking if they have a couple minutes to chat and expressing your emotions to them. Trust me though, you feel so much lighter once it is off your chest. The saying is true, a problem shared, is a problem halved. Interesting insight I learned from all of this is that I am not the only person who has problems. The problems I was having and the problems I face today aren't uncommon. They are quite normal. With every problem comes a blessing and a solution. This is why it is important to understand how you treat people. Everyone is going through something; some are just better at hiding it than others. Be nice to everyone. Be respectful to everyone. If it no longer serves you to have someone in your life, walk away with peace and nothing but good intentions for them. Holding a grudge against someone will only ever negativity impact you.

So, I encourage you to start looking for that trusted person, your vulnerability partner, chances are they will be closer than you think. This is the way to drop your back cover, and when you drop your back cover, what happens to your front cover? It drops too!

Who is the person in the middle then? Who is the real me stuck in-between my front and back cover? How do I find them or even know who they really are? Think back to chapter one, when we defined what success looks like to you? Now, go to your bathroom mirror, or check the background wallpaper on your phone, see your values there? That's the real you. Read your values out loud! Watch yourself in the mirror reading them out! That's who has been stuck in the middle this whole time. Own them and be proud of them. As long as you live your life by your values, lean into your vulnerability partner or partners when you need, and to be there for them when they need you, you'll be amazed at what you'll create in your life.

Breathing is the cure to frustration:

Hands up if you have ever felt frustrated whilst out on your farm at some point. I sure have! I still get frustrated today. Sometimes it gets the better of me, and it is a reality check that I must keep practicing my breathing. We breathe around 22,000 times a day, but it is not something we think about is it. It just naturally happens. We don't sit there and go, oops, I haven't taken a breath in a while, I better take one now. That is because our breathing is subconscious. Similar to our reactions when we become frustrated. The moment we bring breathing to a conscious level, we become more in control.

Let's look at a frustrating situation on your farm. Now everyone's perspective of frustration will be different here. Remember the lens theory we spoke about earlier? Meaning what you deem to be frustrating is a direct correlation of the way you see the world, or the lens you are looking through. Think about a situation you've had on your farm that just tipped you over the edge. It could have ruined your entire day, or if you thought it was really bad, ruin your entire week!

Now, let's question this. Why did it ruin your entire day? Or ruin your entire week? How did you feel after that situation? Maybe drained or exhausted? So, why is that?

It is because you made the CHOICE to react in that way. You saw the situation unfold, then you made a choice to react the way you did. The results of that choice then proceeded to ruin your day or week. As a result of that choice, it made you feel drained or exhausted. So, there was an action or stimulus, and a reaction or response. Bear with me here, there is a cure!

"Between stimulus and response there is a space. In that space is our power to choose our response. In our response lies our growth and our freedom." – Viktor E. Frankl

This is where our conscious breathing technique comes into effect. That gap in the middle. Yes, that tiny little gap in between experiencing that frustrating moment and you blowing your lid. You might not think there is a moment in-between, but there is. This is your choosing moment for your subconscious to kick in and say, 'Hey! I do not like what I am seeing here! GRRRRRR! ANGER! YELLING! SWEAR WORDS! You're up, let's go! All of a sudden, these reactions start coming out. This process could happen in less than one-tenth of a second. What if you could widen that gap? What if you gave yourself more time to think of a reaction. A reaction that will actually serve you positively, rather than destroy you, allowing you to get on with your day in peace and happiness.

BREATHE!

I believe we can all be aware of our frustrations. As humans, we all hold the ability say, that frustrates me, whereas that makes me happy. As farmers, all you have to do is look at every specific job you do on your farm, dive into your memory bank and think back to what occurred whilst doing that job, that resulted in you becoming angry or frustrated. You may have by now fixed that issue so it won't happen again, but I'm sure it is still a memory every time you do that task. So being aware of our frustrations is the first step in this process. I find checking in with myself an easy way to constantly be aware. How am I feeling right now? Not good? Then why? What do I need to change or stop doing to feel good? Or, I am feeling good! That's great! Keep it up! I could do this 20-30 times a day in my head. It also helps me not get caught up in my thoughts.

So, the frustrating moment arises. You're now aware that this is a trigger for you. You must stop what you're doing immediately! Safely, of course. Simply by stopping you're already widening that gap between stimulus and response. I find closing my eyes helps, and I take a deep breath in through the nose, as deep as I can possibly go and hold. One. Two. Three. Let the breath out with a big exhale through the mouth! This relaxes me, and with my eyes still closed I say, 'it's okay'. I have now calmed the situation. The gap is so wide between stimulus and response I could park a semi-truck in there. I have stopped my subconscious from taking over by becoming conscious with my breathing. Is it really that bad? Getting upset or angry won't solve the issue. Yes, this is still frustrating, but is it the end of the world? This can easily be fixed and I can move on. I need to correct this and this, so next time it shouldn't happen. I have total control over my situation now. And every time this occurs, I get better at my response. Let's face it, farming is full of the unknown, similar to life. There is plenty that is out of our control, but there is just as much within our control, within ourselves. So it is wise to use that power of control to your advantage.

I remember a frustrating moment in my first year back on the farm where my reaction did not serve me. I had a load of grain coming in, auger set up, silo opened, everything was looking good. Less than 2 minutes into unloading the truck the auger got blocked and the drive shaft on the auger broke. BANG! I was programmed to get angry. My subconscious kicked into action and said SWEAR! CARRY ON LIKE A PORK CHOP! Do these things and it will make you feel better about the situation. So, I did!

Can you see my front cover in action here? The part about high achiever. The truck driver probably thinks I run my farm with sub-par equipment, but I want him to think I'm a professional farmer with great equipment. In that moment my reaction was more about what I thought the truck driver thought of me, rather than actually working on a solution to my problem. It was interesting looking back on this, I believe the truck driver on some level knew about this stimulus and reaction concept. He came over and said, 'I don't know about you, but I have never seen one of these be fixed by yelling at it.' In the moment, that was not what I needed to hear. however, he was right. What was I achieving by getting angry? Yes, it was a setback. Yes, I did not plan on this happening, but the truth is, it has happened and I cannot change the fact that it has happened. After calming down and resolving the issue, the truck was unloaded and my silo was full. I let my reaction get the better of me on that one. Now, I live by the saying, '<u>it is what it is</u>'. There will always be situations on your farm to get frustrated about. If you focus on your reaction to them, you can become their master.

Chapter Take-aways

1. Your vulnerability is your power, use it wisely.

2. Removing your front and back cover will allow to you live as your true self.

3. A relationship with your vulnerability partner/s is your key to internal freedom.

4. It is okay to say you're not okay.

5. Breathe through your frustrations. Use the breathing exercise outlined in this chapter when you are feeling frustrated.

6. Widen your gap between stimulus and response to become the master of your reactions.

Chapter 4

PRINCIPLE #3: HEALTH IS WEALTH

Enjoying the fruits of your labour

Let me challenge you with this question I'm about to ask. It might make you feel uncomfortable, but I am here to challenge your thinking. Again, please keep an open mind to this, because there is a reason why I am asking it.

At what age are you going die?

Got an answer? No? I didn't think so. No one knows the answer to this question. Honestly, I don't think we would function if we did know. So why do we neglect our health? If you're alive today, that is the biggest and best gift you could ever receive. None of us are guaranteed tomorrow, that's just fact, but we all act as if we are. So why is that? Why do we live today, for tomorrow? This can be quite a hard concept to understand as a farmer. We can all agree that farming is built on delayed gratification. Sow the seeds in autumn for harvest in summer. Feed the animals today for the desired outcome tomorrow, next week or in 6 months' time. That's just the nature of farming. Always has been and always will be. But that's farm-ING. Not the farm-ER. There's a big difference there. Remember earlier I gave you an idea to think about. The idea of, 'you are not your farm'. Think about this for a minute. Are you running your farm? Or is your farm running you? Are you in wake up and go straight to work mode? If you are, then if I am being honest, your farm is running you. Your farm is the master and you are the servant.

Let's touch on delayed gratification for a moment. I am not saying this is a bad thing, but I am saying it can be a trap, especially for farmers, who are surrounded by the concept every day. We have to understand that delayed gratification is the resistance of an immediate pleasure in hope of a more valuable reward in the long-term. Let's look at a farming example here. You receive an amount of money. You could spend it on a family holiday. However, you decide to spend it on the farm, which will improve productivity for the following year, resulting in more money coming in. 12 months pass by, you do the same again. Invest in your farm, forgoing the holiday, but improving productivity, resulting in even more money coming in. 12 months later you do the same again! Can you see the trap here? Let me keep painting this picture for you. Yet another 12 months pass by but guess what!? You do the same again! The delayed gratification theory says that this is going to be one massive holiday. Europe here we come! Unfortunately, no, this is not the case. Look at the habit this has formed. Look at the story you're telling yourself here. 'All money must go back into the farm'. The thought of a holidays most likely doesn't even exist in your mind anymore.

You know, that holiday you could have taken three years ago but made the choice not to?

Now, I'm not saying that reinvesting into your farm is a bad thing, there is definitely benefits to doing that, there is a time for that, and there is a need for that. That is how we grow our farming businesses. However, what about the benefits of investing in yourself? Have you ever thought of that? There is definitely a time for that, and most certainly a need for that. Like investing in your farming business to allow for growth, you must also invest in yourself to allow for growth. So, how much money, time and effort do you invest in yourself?

By you reading this book right now, you're investing time into yourself! What else are you doing for you? I believe investing time in yourself is healthy. Hence the title of this chapter. Your health ultimately creates your wealth. If you had of mentioned the word 'health' to me 5 years ago, I would have cringed! Great, I have to get a gym membership now, or buy some runners so I can go for a run. Unfortunately, at the time, and yes, I'm sure you can guess what I am about to say, I thought this was the answer. So, what did I end up doing, I got some runners and started running. If you have ever seen the movie Forrest Gump, I was a bit like that. I woke up one day and decided to go for a run. I didn't enjoy it at first, most likely due to my fitness level at the time and the soreness I was creating. I pushed through this stage and started to fall in love with it. I got to a point where it became part of my day, it was my thing I did for me. Then came the apple watch. I started to time myself and see how far I was running. Then it became a competition within myself. When you start to look at life as a competition you automatically think, winner and loser. As I was competing with myself, I was ultimately both the winner and the loser. Yesterday's time was quicker than todays, I mustn't be as good today. All of a sudden, the benefits of going for a run was lost and it started to have a negative effect on me. I had lost sight of the whole purpose of this. It was to improve my health! Not to destroy it. I had to find another approach here, I had to find out what health actually means.

My partner was reading a book at the time, 'The happiest Man on Earth' by Eddie Jaku. She said I should read it, so I purchased it online and never gave it another thought. Months went by and I saw it sitting there, so I decided to give it a go. I highly recommend this book if you are in search of happiness. This here was my answer to health that I had been searching for! This is a true story, about surviving the concentration camps during WW2, and how happiness will always prevail.

So, if happiness is superior, how do I make the link between happiness and health? Easy! Happiness is health! Do not think of it as, 'when I'm healthy, then I'll be happy. Think of it as 'I am happy when I focus on my health'. Now, health is a discipline, it requires time and effort. It is simple to do, but also easy not to do. This is where you must make a choice. It may not be easy at the start, there most certainty will be temptations and setbacks along the way, and this isn't something that you can do once and its done. The most successful farmers I have met along my journey so far, all have their own versions of this, but it is something they practice daily, because they know how important it is to them. This should be an exciting new step in your journey. As you experiment with healthy ways that boost your happiness, your life will start to improve! Maybe it is a gym membership, or going for a run like it was for me, that is up to you to decide. The biggest thing is to make that initial decision. 'I commit to improving my health through my happiness'.

'Awesome, now you want me to find more time in the day that I don't have'.

Not necessarily, in fact quite the opposite! How much time in the day do you spend in your Ute? Or on the tractor? Or doing jobs that you could do with your eyes closed? Plenty is my guess! What is the best use of this time whilst performing these tasks? Listening to the radio or listening to an audio book on ways to improve your business or personal life? I have a mentor and great mate who is a dairy farmer. He is one of my vulnerability partners! Every morning it is his job to get the cows in, ready for milking to start at 5 am. Out to the paddock he goes, air pods in and audio book playing. As the cows slowly wander down the lane towards the dairy, off the side by side he gets, around to the front of the vehicle and does 10 push ups off the bull bar. Back in the driver's seat he goes, and moves the cows forward, repeating this process a dozen times all the way down to the dairy. It is 5 am in the morning, and this man has listened to 45 minutes of an audio book and completed 120 push ups! How do you think his day runs after investing this time in himself? Pretty high level and elite if you ask me! This is what I am talking about. He is still performing his routine task on the farm but has made a decision to make the most of his time. He could easily wake up every morning, perform the same job of getting the cows in, maybe complain about the cold, scroll social media whilst they wander down the lane to the dairy, and think about what he is going to get at the bakery for breakfast. Which of these two versions do you think will create success? Sure, the first version isn't easy to do. It takes commitment and discipline, but it has become part of his routine now and it is exciting for him! He looks forward to it every morning. He is investing in his health and happiness!

So, how can you create your version of this on your own farm? I recommend start with doing it on your farm in-between jobs, or whilst performing routine tasks because you're doing them anyway. Get creative with this, be safe and have fun with it. No-one has to know about it if you do not feel comfortable sharing. Make it 'your thing'. Explore the subject of self-help more! There is an abundance of information out there that can help you form routines, or give you ideas on strategies you can implement to see if they work for you. I do know that if you stay committed to this for long enough, there will be a point when you will make the time for your health. A family walk at sunset, waking up an hour earlier to read a book, or exercising with a friend. This will just naturally happen because your body will start to feel great, your mind will start to clear up and you will want more of that for you. The most successful farmers make time for their health and happiness.

I mentioned earlier that this isn't something you do once and it is done. 'The Slight Edge' is a great book, written by Jeff Olsen, who explains it quite well. He says throughout your day you have an endless number of moments where you have two choices. Choosing option #1 adds to your life positively, and choosing option #2 adds to your life negatively. Questioning these moments, and asking 'does this add to my life in a positive or negative way?' Choosing to mind numbingly scroll social media over reading a book will add to your life negatively. Whereas, choosing to read a book over mind numbingly scrolling social media will have a positive affect! Let's be real here, we are all human and we are allowed to treat ourselves. Choosing a healthy meal option 8 times out of 10, with the exception of KFC or a pizza 2 times out of 10 is okay! We just need to be consciously aware that we are feeding our positive side twice as much as our negative side. The results of depositing more positive actions, rather than negative ones, is that the key areas in our lives, such as finances, business, health, relationships and personal develop, all start to reflect positively in our lives. Our lens in which we see these key areas in our lives shows us positive outcomes.

That covers off farm activities, incorporated into your day, that can improve your health. How about off farm. Have you thought about healthy off farm activities before? I called a farmer one day for some advice, but got no answer. Later that night he called me back and said that he had been playing golf. I left it at that, and we moved on with the conversation that I wanted to have. From the conversation around the advice I wanted, he said, when you're free next, come over and I'll show you in person what I am talking about. I never let an opportunity slip when someone invites me to their farm. I love seeing how others are running their operations, especially if they run similar enterprises to me. About a month later, I had a quiet day, so I decided to take him up on his offer. I gave him a call, however, no answer. He returned my call later that afternoon, saying, he was playing golf. My curiosity got the better of me. Why was he playing golf so often? For no other reason apart from satisfying my needs of knowing this farmer's personal life, which was none of my business anyway, I said, 'gee, you really love your golf don't you.' He laughed and replied, 'this is just what I do for my down time.'

I finally got to visit his farm. Driving around the paddocks, talking systems, procedures, ideas and operations, we ended up at his kitchen table. With the conversations flowing I asked, 'so when are you playing golf next?' Thinking I would get a typical answer of, 'when this job is done', or 'once we finish this, then it'll be quiet'. He then pulled out his diary and said, 'next Wednesday'. Wait, what? He has it scheduled into his diary? This wasn't something I was familiar with. A scheduled day off the farm to play golf, he must be mad. I was curious to know more now. I asked why he did this, he said, 'if I do not make time to get off farm, and clear my mind with something I enjoy, I will forever be stuck here.'

Here is a farmer, making time for himself, as part of his day-to-day jobs list. He scheduled it into his diary with no less importance than his sowing dates, or his calving dates. I was amazed by this. He understood the importance of these 'off farm hobby' days. I certainly didn't understand their importance, but I was determined to find out. I find it incredibly fascinating that when you want to know something, and ask, you will find the answer. I had spoken to many successful farmers before, but none of them mentioned the word hobby to me. Turns out, I had never asked! I didn't think this word existed in the farming community. 'I'm too busy for that' or 'when would I get time to do that'. Sound familiar? Maybe you're saying that to yourself right now, as you are reading this. So, I had to go back to my list of successful farmers and ask them, what does the word hobby mean to you? To my surprise, every single one of them had some sort of hobby. They all made time for an off-farm activity, that would relax them, take their mind away from the farm, and allowed them to recharge their batteries.

There was definitely a lot of golf! There was also bike riding, hiking, chess, sports coaching and photography to name a few more. They all said they had to schedule it in, and be disciplined on it, or else it would never happen. Now, there are plenty of unexpected things that come up in a farms day-to-day operations. So, they had flexibility in their hobbies, but they made sure that if it wasn't possible today, they had to make it possible tomorrow. I've done this before. Commit to a day off farm for myself, get home and feel great, then the next time it comes around, I've chosen the farm over the hobby, and soon enough it just disappears out of your mind. This is easily done. It made it easier for me when I chose something that did not interest me. I mean we aren't all golfers! Along with commitment to their hobby, came a very interesting benefit. Their best thinking and problem solving came from when they were off farm. When they were outside of their farming environment, they could clear their minds, relax and solutions would start coming to them. They didn't intend for this to happen, and it certainly wasn't forced, but as they became more present and focused on enjoying their own company, this magic would appear.

Off farm may be outside of your front gates, in a completely new environment. However, I know of a successful farmer who is quite handy on a welder. His hobby is in his shed, turning scrap metal into artwork. This is his definition of off farm. Once he steps foot into that shed, he leaves the farmer behind, and becomes an artist. Unless it is an emergency, no one is to distract the artist from perfecting his craft. He is in the zone, letting his imagination run wild, enjoying his own company and creating masterpieces.

There is a lot of power in this hobby theory. I have seen it first-hand. You may not feel like this is something you can do right now, I thought like this once. I have seen the benefits outweigh the decision to ignore it though. Challenge yourself, if what you have chosen does not feel right to you, make changes and give something else a go. Keep experimenting until you find something that excites you. Something that you wouldn't want to miss out on. Remember to have fun and enjoy this new way of life.

I started this chapter with a very uncomfortable and unanswerable question. I was speaking to someone one day about this topic of health is wealth. I asked them that exact same question I asked you. They looked at me with a very confused face and said, 'well, if I'm not guaranteed tomorrow, why would I bother investing in my health?' This logical thinking farmer had me stumped. I didn't have an answer. Besides arrogantly saying, 'well, why not?' I had nothing. I got back home that night and couldn't sleep. If I was serious about farmers putting their health first, without the guarantee of tomorrow, like I had stated, I need to be able to back it up. I need the farmer to understand the lens I was looking through.

Ture, we are not guaranteed tomorrow, but if we do not prepare for it, and it just so happens to come, what do you think will happen? Tomorrow is only ever less than 24 hours away. And how many tomorrows would you like to have? A lot I would say! I know I want to have as many tomorrows as I possibly can. So, if I focus on my health and happiness, and that increases my chances of having more tomorrows, than sign me up! I want to be around to see my children grow and create their paths in life. I want to see my grandchildren do the same! I intend to also see my great grandchildren! As farmers, I'm sure we all have a great sense of pride when it comes to our farms. Whether it be a multi-generational farm, or you have gone out on your own. So, wouldn't you like to enjoy that for as long as you can? Pretty hard to do it if you do not have your health. Ultimately, the choice is yours, and you're the one that must live with those choices. So, what do you choose?

Chapter Take-aways

1. Do you work for your farm, or does your farm work for you?

2. Investing time and money into yourself is just as important as investing time and money into your farm.

3. Use your time wisely and incorporate positive actions that uplift your health.

4. Question your choices, is this adding to your life in a positive or negative way?

5. Schedule 'off farm hobby' days into your diary and take action on them.

6. How many 'tomorrows' would you like to have?

Chapter 5

PRINCIPLE #4: THE POWER OF NETWORKING

Open up to grow

Farmers are some of the most highly skilled professions in the world. It is not often you see one person wearing a significant number of hats within an organisation, most times all within one day on the farm. The farmer is definitely a jack of all trades. They know their system and their land inside out and back to front. With a highly knowledgeable wealth of information there, I get quite confused when farmers do not engage in openly sharing that knowledge. I perceive it as running a farm is a top-secret mission, you can look but you cannot touch. Or, I will tell you 20% of the information you need but no more. I am not sure where this attitude developed from, and it would be a waste of time finding out. I am more interested in the solution to this problem, and it seems quite simple really. It is 'open up to grow'. This topic can be quite new to most farmers, especially if you have grown up in an environment of keeping all farm details in house. On my quest to find successful farmers, and what makes them successful, I noticed this next theme to be quite common amongst them.

They all share their results to learn, not to brag.

I was lucky enough to meet a group of farmers who were practising exactly this. These successful farmers were all playing in the top 5%. They may not have had a greatest year in their farming journey, but they still wanted to learn from it. Being prepared if a similar year was to come around again, was more important to them than hiding from their results. Likewise, the farmers who had outstanding years wanted to learn how to capitalise more on these opportunities if a similar year was to come around again. Each successful farmer had something constructive to give and got something constructive to take back to their farm. Isn't this incredible? I sure believe it is. Here are these farmers, who have dropped their front covers, and are open to vulnerability. As I sat in on one of their meetings, let me tell you, the amount of knowledge in that room was like nothing I had experienced before. There wasn't any room for ego's either! They analysed each individual farming business, leaving no stone unturned, and asking some pretty hard and tough questions of the owners of these farming businesses. I got the opportunity to have lunch with one of these farmers and the first thing I asked was, 'why do you do this? That seemed pretty brutal to me'.

He looked at me with a smile and said, 'you don't get it do you?'

'Get what?' I replied.

He responded with, 'Nothing ever grows in your comfort zone.'

He said, I look at it this way, I have a lot of respect for the farmers who have decided to do this. From the outside looking in, it's not easy to be ridiculed or questioned. Especially if you have poured your heart and soul into it. But we aren't making it a personal attack. We all go into those meetings with the highest of intentions for the success of the business, and with the highest of intentions for the success of the farmer. No one would be here if they did not want to learn and grow. That is what it comes down to, a non-emotional evaluation of a business. I expect the same from them in return. I want these like-minded farmers to find faults in my business. Things I cannot see because I may be too close to it. None of us are perfect, and none of us have perfect farms. However, each time we get together and do this, we are one step closer! We can all leave that room at the end of the day and can still enjoy each other's company. We check in regularly throughout the year with one another, we genuinely care for one another, we all want to see each other succeed.

That was it, that high level thinking and vulnerability of a successful farmer. I had yet again found another piece of the missing puzzle to success. I just knew I had to get involved! Looking back now, being a part of something like this was the best decision I have ever made. Not only for my business, but for myself too. I could grow my farming business and myself all at the same time. Honestly, it wasn't as scary as I thought it would be either. There was a time to check in and express how I felt with genuine care from other farmers. Then there was a time to get down to the nitty gritty of farm performance. What could I have done differently, why I done something a particular way, what were the decisions around choosing to do it this way, what other farmers in the room had done, their success and learnings from it. It was amazing! All because I chose to be vulnerable and could do it around other vulnerable farmers. See, no front cover.

Notice how I wrote 'success and learnings' not 'success and failures?'

The word FAIL is not embraced enough in farming. Well, life in general, but let's stick to farming for now. No one ever mentions the F word. Fail is such a front cover word. We will do anything not to go near it, avoid it at all costs! Never claim that we have achieve it.

But do you know what fail actually means?

F-irst
A-ttempt
I-n
L-earning.

Fail is just a learning experience. If you embrace it enough, you'll soon come to realise that we are failing all the time. Hence, we are always learning!

Next time something on your farm doesn't go the way you thought it might, the results may not have been what you planned. Choose to embrace it! Say, 'YES! I now have an opportunity to learn from this'

Find a way to get better and ask around. I am sure it has happened to someone else before, and they probably already have found a solution for it. If you do not ask, you will not receive. You may not get your answer straight away, but be okay with continuously asking for help. The other side to this story is that you shut the opportunity out altogether, and keep repeating the same thing you've done that got you there in the first place. You know what that is called? "Insanity - doing the same thing over and over and expecting a different result." Albert Einstein

Finding a group of farmers who are willing to share for the greater good of themselves, and their businesses can be tough, but it is not impossible. I guarantee you, the moment you decide that this is something you are willing to do, you will find your group. There are plenty of successful farmers all around the world. They are successful because they chose to be and done something about it. So, if you choose to be successful, then start to do something about it today. Search it online, put the feelers out in the network of farmers you already have, be courageous and step out of your comfort zone.

I mentioned earlier that you're the product of the 5 people you hang around most. If the people you hang around head to the pub most nights after work, guess where you end up? That's right, at the pub. However, if the people you hang around most are creative, problem solve, open, honest and focused on growth, then guess what your chances are of becoming those things? It's a sure thing! Either you become them, or you out grow them and leave. There are no other options. So, what kind of people do you associate with the most? What are the conversations like with them? What is the language that they use? Are they optimistic or pessimistic? Once you understand this concept, it is interesting as you start to question what you are hearing and seeing around you. Now, do not take this literally. I am definitely not saying to assess everyone around you, and who ever does not meet your criteria of 'positive influential people', you must remove them from your life. Please do not do that. This is about becoming aware of the people who influence you. More so, the people who you LET influence you. You can still hang around your mates and talk about the Football, Rugby or soccer. No less than you did previously. It is about becoming more aware of the way they see the world, the lens they are seeing it through. Your job is to take your new found lens, remember the one we talked about in chapter 2, and find the people with that same lens. They are the ones you want to be associating with on a higher level. A level of growth. Let them be the positive influence you are looking for in your life.

Mentors:

Mentors were another powerful tool that all the successful farmers I met have used. I noticed there were quite a large range of different mentors. Not only other farmers, but successful business owners, or wise people outside of farming. These were all people who are experienced and trusted to the individual farmer. I now have four mentors that I personally go to for advice and to brainstorm ideas with. Two are farmers, and two have no real interest in farming, but are high level business owners in their own right. This is a very powerful tool to have in your back pocket as a farmer. The ability to communicate ideas into words, and be questioned on them. To see if they are of any value. This is one of the quickest ways to grow as a farmer. I've had plenty of ideas that my mentors have loved, and also plenty that they have questioned me on. This is the beauty of a mentor. They are not there to answer your questions. Remove the idea that mentors hold the silver bullet. They are there as a guide. They will guide you to the answers you are looking for. So, I challenge you to find your mentors. When I first started mentor searching, this was once I had discovered my real self and had defined my own success. I looked for people who I thought were successful and aligned to my values. I had to be vulnerable here, and simply ask them if they were willing to sit down and have a chat with me. Interestingly enough, the ones who agreed are still my mentors today. Where-as the ones to declined, I still respect them, and still believe they're successful in what they do, but do not fit in as a mentor in my life. Most of them were shocked when I asked if I could consider them as a mentor, as they felt they didn't have much to offer. Maybe I could see something that they couldn't. But deep down I felt that these people were the ones that would help me grow. Remember to be open to this new idea, this isn't set in concrete, if you feel someone has got you as far as they can, find another mentor who will take you to the next level. This just means that you are growing! Either you out grow them, or you stay at their level, and either is fine, as long as you are happy with the level, you're at.

I have an amazing mentor that I inspire to be like one day. He and his wife run a farm, and have two very successful off farm businesses. Now him and I may only talk, on a mentor to mentee level 2-3 times a year. But that is all I need. The wisdom that he shares with me is priceless! The reason I am telling you about him is because I want to share with you, his mentors. How he sees his mentors and the roles they play for him. In his office he has two spare chairs, each one in a corner of the room. This is where his mentors sit. Now, his mentors may not be there physically, but when he is in his office, creating a new idea, or solving a problem, he turns to them as asks, what would they do. His mentors are so powerful and wise to him, that he pictures them in that situation and how they would solve it. You may have mentor's past, or present, who cannot be there all the time, so this could be a great way for you to have them involved. This is about being creative, the place where growth comes from. The choice is yours.

A Coach:

The most successful sports teams in the world have a coach. The most successful athletes in the world have a coach. The most successful business owners in the world have a coach. So, why don't you have a coach? Let's look at this another way. Usain Bolt didn't first become an Olympic champion, then decide to hire a coach. He had the coach first, then came the success of becoming one of the world's greatest sprinters. Do you know what it takes to become one of the world's greatest sprinters? Commitment, discipline and belief. I'm sure there were plenty of times along his journey where he would have rather stayed in bed, instead of going to a run, hitting the gym, or plunging into an ice bath. I know along my own journey; I've had mornings where I would rather stay in bed. Especially on those cold, wet, rainy mornings. Bed is the easy option, but is it the right option? It could possibly be. Your body may be telling you to rest. I think it is important to listen to your body, however, when we listen to our mind, and it is saying, 'stay in bed where it's nice and warm', that is coming from our subconscious. Remember, our subconscious will do anything to stay in comfort, this is called the comfort zone. So, you're saying I need a coach to get me out of bed every morning? Definitely not, but if we look at the role of a coach, it is easy to see why you would get out of bed.

Every time an athlete or sports team win a significant title, they may thank the coach, but who is the one receiving the medal or the trophy? Who is the one that is written into the history books of time? It is the individual or team who was coached. The coach is rarely mentioned. Majority of you reading this could tell me that Usain Bolt did in fact win gold at the Olympics, but would struggle to tell me who he was coached by. That's because the role of the coach isn't about receiving awards, their reward is in them getting the best out of the individual or team. I'm sure it would be quite satisfying to have coached an Olympic athlete, a great accomplishment to add to their CV. However, that is a bi-product of them becoming the best versions of themselves.

The coach is there to hold you accountable to what you actually want. They do not set the direction, you decide all of that, because this is your life, isn't it? If you are lost with your direction, that's okay, we will dive into personal goals later on, and I will guide you through how to set up your direction from there. For now, this is an insight into what I have seen the most successful farmers implement. That being, a coach.

Let's look at this idea for a farmer. A farmer is a business owner, right? Well, the most successful farmers I have met consider themselves business owners. So, if the farmer is both the owner of the business and works within that business, who is he or she accountable to? If something doesn't get done, they don't have to report to the boss and explain themselves. They are the boss! If this was the case, they'd have to stand infront of a mirror, and I don't see that happening. Sure, you may be accountable to external factors, as in repayments, but I am talking about personal accountability. What happens when someone is unaccountable? Say you organise to meet up with a great friend. You have a great relationship with them. Together you decide on a time and location. You restructure your day so you can meet up with them. You arrive at the location, at the agreed upon time, however, they aren't there. You give them the benefit of the doubt, they may just be running late. Half an hour goes by, you decide to message them, and they reply saying, 'oh no! I'm so sorry, I forgot all about it.' What happens to your relationship with them? Probably not much. You may be annoyed but when you see them next, you'll laugh it off and move on. However, this same person, does it to you again. Another planned catch up, you restructured your day to fit this in, but they go and pull out on you at the last minute. What happens to your relationship with them now? This is the second time this has happened. You may be a little more annoyed this time. There's probably some doubt creeping into your mind about their commitment towards you. But hey, they're your best friend, and best friends let each other off the hook from time to time. No point ruining a good friendship by starting an argument. Weeks go by and they send you a message, asking if you would like to catch up. You agree, but now you're somewhat cautious. You let them pick the time and location that best suits them. You reconfirm with them a couple of days prior, to make sure they are still right for the catch up. You message them when you are leaving home so they know you are on your way. You message them when you get there so they know to look out for you. Then they send you a message saying, 'running late, can we make it for an hour's time?' Can you see what is happening here? There may well be genuine reasons as to why this person isn't committed, that's not the point though. Can you see what is happening to you? The person who is being let down. Look at how you see this relationship now, compared to how you saw the relationship previous to these events. What story have you created in your mind about this relationship now? If they were to message you again and want to catch up, how would you feel? What would you say? You've been let down by this person multiple times before, would you go out of your way to see them again? Or would you make them come and see you instead? You probably have little faith in them now. You question their commitment towards you, maybe even make a promise to yourself not to go out of your way for them anymore? 'If they want to see me, they can make the effort this time.' The relationship has gone from great friends, to you feeling unwanted, so you give up on it.

Now, let's replace you in this story, with your mind, and you now become the friend that is uncommitted. Let's run this scenario again and see how it plays out. You say to yourself, 'I am going to get up early tomorrow, and have all my office work done before 10 am.' You've now made a commitment in your mind. Tomorrow morning comes around, your alarm starts to ring, you turn it off, roll over and go back to sleep. Opening your eyes, it is now 9:30 am, and you've completely forgot that you were meant to be up, and have almost all of your office work completed by now. That's okay, you let yourself off the hook. It was one time, no big deal. A few days pass by and you say to yourself 'I am going to go to the gym tonight.' You're keen and excited to go. On the way to the gym, you get a phone call. It's your neighbour. 'Gidday mate, I've finished early today, are you keen to go to the pub?' You agree, and head to the pub instead of going to the gym. What is the relationship between you and your mind like right now? You've made two commitments to yourself in the past few days and you have not followed through with either. There may not be any consequences as yet. Remember 'The Slight Edge' theory from last chapter? Is breaking a commitment to yourself adding to your life in a positive or negative way? Later that week you say to yourself 'as soon as I get home tonight, I am going to begin reading that book.' You finish work for the day, head home and pick up that book, but you remember the cricket is on! There is only an hour left of the cricket, so you say 'I'll watch that, then I'll start that book'. See what is happening? Put yourself in the shoes of your mind. Your mind is probably saying, 'here we go again, old Mr. commitment breaking yet another promise he has made with me.' So, what happens to your mind now? It starts to doubt you. It starts to have little faith in you. It starts to question your commitment. It starts to feel unwanted. This is all happening on a subconscious level, so it won't be clear at first that this is what is happening. Basically, you are writing a program. You're the one adding fuel to the fire of doubt, little faith and poor commitment in your mind. Over time, as it is now a subconscious belief, you will begin to doubt yourself. You will struggle to commit to things because you have little faith in yourself that you will follow through with it. Your confidence will be destroyed, and when your confidence is little to none, you begin to feel unworthy. This is why I am very aware of what I commit to with myself. I am aware of what I say to myself. If I say I am waking up early, I make sure I follow through and do it. No if, buts, or what. My commitment to myself is my number one priority. Really challenge yourself on this one. Your relationship with your mind is so strong, and it should be respected.

So, what does a coach have to do with this? Well, this is what my coach taught me. They taught the value of commitment. They made sure I followed through with everything I said I was going to do. They kept me accountable and taught me how to be accountable. At the time, I had no one to be accountable to. No one was going to ring me if I turned up half an hour late. No one was going to argue with me if I didn't make certain decisions on time. I learnt the lesson on how not to over commit and under deliver. My coach knew if I were to do these things, that I would get to where I wanted to go. Do you want to know the scary part? Paying a coach, who you talk to once a month, and telling them you didn't do anything from the previous month's commitments. I learnt that straight up in my second ever coaching call. If you want to waste your time and money, go to a casino was the response I got. Never again did I let my commitments slip. A coach should be firm but reasonable. Your coach is not your friend. I have a great relationship with my coach, but when it comes to coaching, her nice hat comes off, and her coaching hat goes on. She knows me, she understands me, and knows what I need to do to push myself, stretch my thinking, and be the best version of myself that I can possibly be.

Networking is all about the people. As a farmer, I encourage you to network with not only other farmers, but people from all different backgrounds. You can learn so much from doing this, they can learn just as much from you also! It is all about different perspectives in life. The more you spread your wings, the more open you are to new ideas, the more you allow yourself to grow. There is so much more beyond the boundary fences of your farm, it is all there just waiting for you to experience it. If the most successful farmers I have met, all have a high-power networking group surrounding them, my question to you is, 'who is in your network? What new network do you need to form to start moving towards your success?'

Chapter Take-aways

1. Successful farmers, the ones playing in the top 5%, share their results to learn, not to brag.

2. There is no growth in your comfort zone.

3. F.A.I.L to a successful farmer only has one meaning, First, Attempt, In, Learning.

4. Who are the 5 people you surround yourself with and let influence you?

5. Be courageous, go out and find great mentors who will expand your growth.

6. A coach is a powerful accountability tool to use not only in your personal life, but in your business also.

Chapter 6

PRINCIPLE #5: GRATITUDE

The secret that changes everything

Where do you focus most of your attention? Is it on what you want? Or what you already have? I could ask any farmer, if they could snap their fingers and have anything magically appear, what would it be? The typical response being, 'well, how long is a piece of string?'

The farmers wish list is endless! How often do you drive around your farm with your wishful thinking glasses on? I really want a new tractor. A new shed there would be nice. Some new fences along the road would make the farm look neater. Sounds familiar? How about this one, 'I wish I had more money to do those things.' These are quite common in agriculture. Wanting these things is not a bad thing, but losing sight of what you already have, or what you have achieved, in order to chase the shiny object, can be a trap that I recommend you avoid.

I think back to my first year on the farm, and one thing that shocked me the most, that I certainly wasn't ready for, was that how slow farming was. I saw farming as a play and wait game. Make your move, wait, see the results, readjust, wait your turn, make your new move and repeat. Maybe I was looking at it wrong, but that was how I felt at the time. Then, in conversations with the older generation, they start telling me how quickly time goes. One season leads into the next and before you know it, you have thirty seasons under your belt. I wanted to push production though, have this done and have that done by these specific times. What I was seeing versus what I was being told were two completely different things. My lens was different to theirs. I knew what I didn't want. I didn't want to get to 60 or 70 years old, saying 'I wish I had of done this', or 'I should have done that'. So, I knew what I didn't want, all I needed to do was to turn that into what I did want. Then find the answers to it. 'I want to look back on my farming career without any regrets, and with pride in everything that I have achieved'. There was my new story I started telling myself.

Finding the answer to this came in a very unusual way. I was sitting on the tractor one day, listening to a podcast. See, I was making good use of my time and investing in myself. Listening away, a special guest comes on. Hayley Grosser – The Abundant Farmer. You can find more out about what Hayley does at – hayleygrosser.com

As I was listening my mind was racing! What was she on about? I couldn't quite understand what she was saying, but she had my full attention. Confused as I was, I kept pushing through, making notes in my phone of everything I needed to understand more about. There was one word in particular that she kept on using, but I had no clue as to why. The word was 'gratitude'.

I couldn't help myself; I needed to know more. I needed to understand what she was talking about. I tracked down Hayley as soon as the podcast was over and sent her a message. See, again, dropping my front cover and asking for help. This gratitude thing, why is it so important? Why are you talking about it on a farming podcast? Maybe I was naïve, but I know I was definitely curious! It is amazing what you learn when you can admit that you don't know everything. As the saying goes; be the dumbest person in the room, and if you are the smartest, you're in the wrong room.

Gratitude is the act of seeing what is, rather than seeing what isn't. Once practising gratitude, you can then be in a state of gratefulness, which is a state of love. Love is the highest power a human can possess. It is a weird feeling to walk around your farm and practice gratitude. The first time I done this, I went around saying things like, 'I am grateful for this grass I have planted for my stock to eat', and 'I am grateful for this rain for my crops to grow'. That's what I thought I had to do. But there is more to this, and it took me about two years to finally understand it. As I said earlier, words are powerful things, but add feelings into those words, and this is where the magic really starts to happen!

I am sure we have all experienced the negative side of this before. Have you ever had a situation unfold in your life that wasn't pleasant? That situation plays over and over in your mind. This could go on for days, weeks or even months after the event occurred. In your mind you think of the best comes backs to say, or what you would do differently if this exact situation happened again. Am I right? Now, the chances of that situation happening again may be slim, but you still play it over and over in your mind. As you play this scenario out in your mind, what is happening to your body? Is your heart rate rising? Adrenaline starts pumping through-out your body? Yes! This is because your mind does not know the difference between a thought, and real life. If it is playing in your mind, your mind now thinks that it is actually happening so it prepares your body for fight or flight mode. Remember this is not a healthy thing to do. Replaying past events, or negative future events that haven't happened, in your mind that you have no control over, is only torturing yourself. Reviewing past events, understanding better outcomes and learnings, then letting go of them is the best practise you could do. If you need help with this, reach out to your vulnerability partner, a problem shared is a problem halved, or visit the resources I have provided at the start of chapter 3.

So, if that is what we can do with our minds, and we find it so easy to do it in a negative way, why do we find it challenging to do it in a positive way? In my opinion, I feel as humans, we are either thinking into the future, or reflecting on the past way too much. We rarely find ourselves in the present.

"The past is history; The future is a mystery; This moment is a gift; That is why this moment is called the present; Enjoy it". – Alan Johnson

Becoming present:

Remember the breathing technique I spoke about earlier? The one to use when you feel frustrated? Well, that same technique can be used here to become present! The idea of being present is all about your minds talk. The average person can have around 60,000 thoughts a day. That's 60,000 little conversations going on in your mind. One after another, nonstop. But what if we could give our mind a break? A rest from all these conversations for just one moment. Well, with breathing and gratitude we can! You've probably heard of the word 'meditation' before. Immediately, you may think of a monk, or someone sitting cross legged, eyes closed, with their thumbs and index fingers forming a circle, repeating the word 'ohm mm'. That's the exact image I that popped into my head when I first heard that word. Like in the movies! However, it doesn't have to be like that! Meditating is simply 'focusing the mind'. We all hold the ability to focus our minds. This is one of those practises that is simple to do, but also easy not to do. It's so easy to always be in a rush on the farm, one job complete, onto the next, and so on. What if you stopped for just three minutes a day? Three minutes isn't that much time. I'm sure you'll still get the same amount of work done. What do you have to lose by giving it a go? If you feel it's not for you right now, that's okay too. However, I want to challenge your thinking on this one.

Why do you think a personal trainer schedules rest days into a training program? Could it be a strategy that allows them to take on more clients? Or could it be they understand the importance of rest? More than likely, it's the second option. Resting in a training program helps you avoid injury, sleep better, promotes muscle growth, and enhances your overall recovery. So, let's look at this as a training program for your mind. How much rest are you giving your mind right now? If all it takes is three minutes per day, and the results will allow you to avoid injury, sleep better, promote muscle growth and enhances your overall recovery, wouldn't you be a little curious to give it a go? I don't mean a physical injury here, although avoiding this could become a result. I mean if you're thinking more clearly throughout the day, avoiding an injury in your mind, as in uncontrollable negative thoughts compounding. Likewise, promoting growth is promoting positive thoughts, and enhancing your overall recovery, think of it as how you react to life and farm challenges.

To practice the art of gratitude, you firstly need to be in a present state. I find it easier to be alone and in a quiet place to begin with. Find that place on your farm that is peaceful for you. The tree at the top of the hill, down at the creek, or in the middle of your crop. Where ever it is for you, find that place. Now, you're in that place, get comfortable, sit or stand, close your eyes, take a deep breath in through the nose and hold, one, two, three. Exhale out through your mouth. Repeat this as many times as you need to, in order to feel relaxed. Feel your breath going in, filling your lungs. Feel your breath going out. As your breath leaves your body, feel your body relax. What can you hear around you? Listen to the sounds, focus in on them. The birds singing, or the trees rustling in the wind. Identify as many sounds as you can. Feel the wind on your face, feel the temperature of the wind, feel your body relax. You're practising being present. Be in this state for as long as you feel necessary. Once you open your eyes, and they adjust to the light, the first thing you see, I want to you find one thing you're grateful for about it. Say it out loud. 'I am grateful for the water in this dam that allows my animals to stay hydrated'. Keep looking! What else do you see that you're grateful for? You're now practicing gratitude, in a present state. The more you do this, the easier it will become. Practice and consistency are the keys here. Soon enough you'll be able to do it from your Ute as you drive around the farm. From the tractor window as you go up and down the paddock. From the moment you wake up in the morning. This may all seem a little strange, I get it. 'Farmers don't do this stuff'. Well, I want to let you in on a little secret, I've done a lot of digging around on the topic of gratitude, and guess what. All successful farmers practise gratitude.

Think back to the start of this book when I spoke about the two very similar farms, with the exact same enterprises, with almost identical results, having two completely different reactions to those outcomes. I couldn't understand why at the time. If you haven't notice already, I was the farmer with the negative reaction, whereas my counterpart was happy, relaxed, and had a sense of joy around what they had achieved. The missing piece of the puzzling for me, but what they had already discovered, was gratitude. When this became clear to me, I had to find out what gave this farmer happiness, relaxation and joy. Again, by dropping my front cover and being vulnerable, I could finally understand why. He said, 'look around, look at what I get to wake up to every morning. Look at the peace and quiet surrounding us right now. Look at the open spaces, the incredible crops out my window, the lambs all running around playing, the cow sitting under the trees. How could I not find happiness in this?' That's when it hit me! FIND THE HAPPINESS!

When you start to look at your surroundings with joy and happiness, your perspective on life changes positively. I can fully understand the rollercoaster effect in life and in particular farming. You may have the highest of highs when life feels amazing, everything is just working out perfectly. Then there are the lows. You're constantly tired, in reactive mode, just thinking about making it to the end of the day so you can sit inside, put the tv on and forget about the farm for a while. This happens, that's just life. If you do not experience the lows, you will never be able to appreciate the highs. Have you ever found that your lows come from external sources? The price received for your product, or the weather reporter saying there's no rain forecast for the next month? I know this is where my lows came from. I was attached to the outcomes of certain things because that's where I thought success come from. I had to know more about the lows. It's all well and good for these successful farmers to say 'find your happiness', but what if happiness feels miles away? How do I find it then?

My love of questioning may be off putting to some, but truly successful people enjoy being questioned. It makes them think, and helps them to grow too. So, if finding your happiness and practising gratitude is something all these successful farmers do daily, what do they do when things seem tough, or aren't going the way they want? The responses I received from this question were very mixed, but upon analysis, they had three very common themes to them. Individually, each farmer I asked had their own special way to overcome their lows, just like they have their own special ways of running their farms, practising their gratitude and being present. The three common themes I found were as followed; Proactive, Self-fulfilment and Choices.

Proactive:

The successful farmers were all proactive towards the current environment in which they are involved in. Being proactive is the action of controlling a situation rather than just responding to it, after it has happened. They all have plans and strategies in place. If that does or doesn't happen by this date, then this is the plan and this is the strategy we will use to execute that plan. They all knew what they were going to do, no matter what the season threw at them. They were all comfortable knowing exactly what the results would be in any situation. Look at how these successful farmers could control a situation that they have absolutely no control over. My question to you is, how are you being proactive to lessen the likelihood of being caught in a low due to circumstances beyond your control?

Self-Fulfillment:

Self-fulfillment for all of these successful farmers came from something bigger than themselves. They did not attach their self-worth to their crop yields, or their lambing percentages. They understood the true meaning of 'you are not your farm'. Their self-fulfilment came from their values. The exact ones that you defined for yourself at the start of this book. They understood that as long as they were living true to their values, they could be at peace within themselves. No matter what situation came their way, their values were their highest priority. My question to you is, where do you attach your self-worth, and what changes do you need to make, to improve where your self-worth comes from?

Lastly, all of these successful farmers made the choice to choose positive over negative. They chose not to listen to the negative talk at the pub. They chose not to listen to the 6 o'clock news. They chose not to be inundated with ag media scare tactics. Sure, they all stayed informed with current events, market trends and business outlook, but they did not let this affect them. They all make the choice to be optimistic. They understand where they must get their information from.

They understand the difference between fact and opinion. They do not allow emotions to override strategical business decisions. My question to you is, what choices are you making? Are they having a positive or negative affect on you? Where is your information coming from? Is this information a fact or an opinion?

All these successful farmers could agree that they still experience low times. At the end of the day, they are only human. However, they all had the awareness of being in a low state, and all had a strategy in place for them this occurred. Some had to get off farm for the day, while others just had to go inside and be with their partner and/or children. This strategy will be different for everyone, and may change with different circumstances. The key here was to know your strategies. The simplest way I had it explained to me was by answering this question. 'What do I need to do to turn this feeling around?' For me, now, all I have to do is force a smile onto my face. Give it a go next time you're feeling down. Close your eyes and smile. You may start to laugh a little. Happy memories will start coming into your mind. Do you know what is happening here? It is impossible to think negative thoughts and be happy at the same time. There-fore, if you're happy, you do not have the ability to think negative thoughts, only positive one. Even though you had to force the smile onto your face, your body automatically associates a smile with positivity. Remember, your mind does not know the difference between real life and imagination. The snowball effects of this forced smile soon start rolling into happy memories, one after another. You have just changed your bodies vibration. You have gone from low-to-high in less than a minute. It is not about neglecting our low times, they are there for a reason, and we definitely do not want to ignore them. Trust me, you do not want to start the habit of filling your emotional filing cabinet. We want to embrace them, thank them, understand where they are coming from and be okay with them. We are all doing our best, and that's all we can ask of ourselves. Although, you have two choices here. Either you to stay in the lows/downs, and let them identify you, or you can embrace them, and choose to make a change. The choice is up to you.

Another technique I use, that is simple and affective, is a gratitude coin. I carry around a single coin in my pocket. This is my gratitude reminder. Every time I put my hand in my pocket and touch the coin, or feel the coin on my leg, it triggers my gratitude reflexes. Where ever I am, whatever I am doing, I have to just find one thing to be grateful for in that exact moment. A quick and simple practise. It only takes 5 seconds to perform. Think of how many times a day you could practise gratitude by simply doing this! I look at it the same as carrying a dumbbell around. Every time I practise gratitude, is the same as doing one rep with a dumbbell. However, instead of building the muscles in my arm, I am building the gratitude muscle in my mind. The stronger this gets, the happier you will become.

We are so lucky as farmers, we are surrounded by pure joy on our farms every day, and we should be proud that we get to do this for a living. So, take the time to find the happiness on your farm today. Look for the little things to be grateful for. Appreciate all that you have achieved, where you've come from, where you are today, the sunshine and the rain, find as much happiness as you possibly can, and remember, what you appreciate, appreciates.

Chapter Take-aways

1. Gratitude is the act of seeing what is, rather than seeing what isn't.

2. Replaying negative past or made-up future events in your mind is only torturing yourself. Let them go.

3. All it takes is three minutes a day to begin with. Use the three-minute exercise outline in this chapter to stop, to practice becoming present and to practice gratitude.

4. Use a tool like the gratitude coin to grow your gratitude muscle daily.

5. When you start to look at your surroundings with joy and happiness, your perspective on life changes positively.

6. Become a proactive farmer where you can control a situation rather than just responding to it.

7. Detach your self-worth from your farming results and attach it to your values.

Chapter 7

PRINCIPLE #6: FOCUS

"Energy flows where attention goes"
– Tony Robins

Have you ever heard the saying, 'if you chase two rabbits, you will not catch either one?' I first heard this back in 2019 whilst at a farming conference in Adelaide, Australia. At the time, it went in one ear, and out the other. I didn't think much of it at all, until I was on my quest to discover farmers success. When I was back home, finding my feet in this game called farming, I was always coming up with new ideas. Each day it seemed that my plans would change. There was a new big idea, I'd start to implement it, but then I'd get bored, or change my mind. I'd think, I need something new and exciting! Then I would wonder why my results were so poor. You cannot expect results through incompletion! I now know this to be true, but at the time all I wanted to be was the next Elon Musk or the next Jeff Bezos of farming! See, back then, my success was money. If I had lots of it, then I'd have lots of success. I am now glad I have discovered a new version of success, but staying true to my word, this book is about my journey to find success as a farmer. I first had to have that version of success to discover my new and improved version.

Throughout my journey, talking to an abundant number of successful farmers, I noticed something very interesting. When I would talk to a cattle farmer, they could speak about cattle for days on end. Likewise, when I would talk to a broad acre cropping farming, the same would happen again. They could talk about crops, tractors, harvesters, yields and everything else that goes with cropping, until they were red in the face. They lived and breathed their farming system of choice. A light bulb came on in my head one day. I was talking to a farmer who had a large cropping enterprise, which was successfully complimented with a sheep enterprise. Driving around his farm one day, it was all cropping this, and cropping that. Not one mention of sheep, which intrigued me. To be honest, the sheep interested me more. They ran a very profitable business, so why wasn't he including that into our conversation? It eventually got the better of me and I just had to ask. 'You've spent the last 2 hours talking about nothing but crops, I thought you were a crop and sheep farmer?'

He said, 'that's right, we are, but when I am out here, I am not the owner, I am the cropping manager. This is where I do my best work. If you want to know about the sheep, you'll have to talk to my sheep manager.'

I couldn't argue with that answer, but I had to ask why. I couldn't go home without knowing why. He took me into his office and showed me the answer I was looking for. FOCUS.

The farm itself had a core focus. It was focused on growing crops and growing wool, and allowing the mix of both, to complement each other. The crops had its own focus, and likewise, the sheep had its own focus. He said to me, if I were to focus a third of my attention on the farm business itself, a third on the cropping enterprise, and the other third on the sheep enterprise, what kind of job do you think I would do? I had never thought of it this way before. This successful farmer was giving his full attention to only one particular part within the business, whilst having full trust in his manager to hit certain KPI's with the other part of the business. Although I had a huge insight into this, I still wasn't satisfied. What about a farmer who works by him or herself? How do they stay focused? I had to now find a successful farmer that fit these criteria.

I have joined many different short-term farm-based courses along my journey. With the intention to improve my on-farm management skills. The beauty about these courses, is that you get to meet a variety of different farmers, from a variety of different areas. We were out on farm this particular day, and the farmer was explaining their operation and what they do. She then went and said, I am the only person on farm. I had regarded her farm as quite successful, so I had to question her more, how does she stay focused with so much going on?

I think she was blown away by my interest in the farms operation and how she managed it by herself. The way it was ran was no different to the previous farm. Focus on where you do your best work, and ask for help, or schedule time to do the rest. I got the full insight into her business and was satisfied that focus was a key principle to farmers success.

She made it very clear to me, that her genius zone was wool production. A 'genius zone' is the place where your passion drives from. It is where you do your best work. Where you could do that work and happily be unpaid whilst doing it for the rest of your life. I believe as farmers, we all have a 'genius zone' within us. Her passion for the animal was elite. All of her focus and attention were on her farming system. 'I grow grass and convert that grass into wool for the lowest cost possible'. I wanted to know more about the other roles farmers have to play within their businesses. I asked her how she felt about being in the office. Especially when it is tempting to go outside and find something else to do. She said she use to struggle a lot with the idea of being inside in office, but had found a way to overcome that challenge.

Within her business she knew what she loved to do. The things that really drove her passion and joy. Her genius zone. She also knew the things that she didn't enjoy doing. The jobs she'd put off or avoid doing until the last minute. At the start of her career, she had to do it all. After all, she was the only one in the business. However, as her business grew, so did her knowledge, which led her to discover the phrase, *'delegate to elevate'*.

This changed the way she viewed her business, and eventually resulted in creating more profit. She understood that if she were to focus on what she loved doing, and delegated the tasks that she didn't enjoy doing, her business could have the potential to grow. This didn't happen overnight, however, over a period of around 24 months, she started to slowly let go of control of the tasks she dreaded doing. Her number one most dreaded task was coding invoices into her accounting software. By employing a virtual assistant, who came into her business initially for four hours per month, she essentially freed up four hours of her time to focus on her genius zone. Once she could see the benefits of letting go, the snow ball effect started to take action. Once she was happy with her tasks within the office were being done to her standard, she would let go of more, and allow the virtual assistant to do more. She then moved on to look outside. Out to the paddocks. Bringing in contractors for animal health care, tractor work and fencing. In just 2 years, her business has gone from her doing everything, to now only focusing on her passion. She still sits in the role of CEO for her business. She still has full control of decision making and financials. However, the time she has freed up by letting go of tasks, has allowed her to make better strategical decisions for her business. Decisions such as pasture species, grazing management and animal genetics.

This can be confronting for farmers to hear at the start. There is a saying that makes me laugh every time I hear it. 'If you want something done right, then you have to do it yourself'. Let me ask you this though. Do you think Elon Musk cleans his own house? Or Jeff Bezos mows his own lawns? If I had to take a calculated guess here, I would say the chances that they do these things, are very low. Why is that? Probably because they are both billionaires and can afford it. Well, they are definitely both billionaires, and hiring someone to do these tasks wouldn't necessarily hurt their pockets. I think there is another reason though, something they know that most people struggle to understand. It is the art of hiring someone who can do the job better than you can. To give them the tools they need to perform that specific job, then get out of their way.

Andrew Roberts, co-founder of Farm Owners Academy, told me something very powerful when I was first learning the art of delegation. He said, 'you will never get rich if you mow your own lawns or clean your own house'. He said you have to look at it like this. When you are in your genius zone, as a general rule, that is around a $500 to $1000 an hour task for you. For example, evaluating animal genetics for the future of your business verses coding invoices on your accounting software. Both very important to your business but have two very different values in your business. If genetic gain can potentially bring in an extra $50,000 to your business over the next three years, and you can pay someone $30 an hour to code invoices for you. It is easy to see where your attention needs to be focused. As farmers and business owners, it is important to value your time and see where it is best spent for the greatest return. We all have a limited amount of time, use it wisely.

Now, mowing lawns or cleaning may be something you love doing. That's okay. I am not saying that they are the only two things that are potentially holding you back. This idea is to challenge your thinking. Assess where you focus your time today. Are there any low value tasks that you could pay someone else to do? Essentially it is about you buying back your time. I encourage you to explore this idea. If it doesn't work out, who cares, you would have lost nothing in giving it a go.

You can only focus on one thing at a time, right? Do you ever catch yourself watching a tv show, and being on your phone at the same time, then not being able to remember what happened on your show, or what you were looking at on your phone? Interesting isn't it. There's that two rabbits' theory playing out in real life. This can be the same on your farm. If you have multiple jobs going on at once, what will the standard of their completion look like?

The most successful farmers I have met all believe in this focus theory. They could all show me what each member of their team were focused on, and how it all came together to achieve the farms focus. This is the farms focus though, but this book is based on the farm-ER, for the farm-ER. So where does focus fit in there? Funnily enough, these successful farms that were driven by focus, were only a bi-product of the farmers focus on themselves! All these farmers that I have met, understood focus on a personal level first, then they just so happened to leak it out into their businesses.

These successful farmers were so focused on their personal goals and values, that there was no doubt in their mind that they would achieve them. They were all focused on winning. This wasn't to achieve a material outcome, but to achieve self-fulfilment. The material outcomes are an added bonus, not the objective. If they do become the objective, you then start to lose your true self. You forego your values in exchange of chasing these material things. This is a terrible trap, and one I urge you to avoid.

The successful farmer knows their values, the ones they wish to live their life by. They then have personal goals written down that align with their values. Goals they wish to achieve in the next 12-months, 5-years, 7-years, 10-years, and even 20-years for some! The 20-year goal was big enough that it was almost scary to think of its achievement. However, if it were to be achieved, it would be an amazing triumph. I saw a farmers 20-year goal once that was to multiply the size of their business by 10. Taking them from 50,000 acres, right up to 500,000 acres. The 20-year goal is summed up by a quote from Norman Vincent Peale, who said. *"Shoot for the moon. Even if you miss, you'll land among the stars."*

These farmers all set an end date to their goals, because a goal without an end date is just a dream. These goals had one specific purpose. When I achieve this, that will increase my ability to live by this value. Their focus is on their true self. When I first started goal setting, mine were all about material possessions. My goal is to have this much land, this type of tractor, get these results, buy this, buy that, have this amount of money and live happily ever after. I hadn't fully understood the purpose of goal setting. Don't get me wrong, these were all things I wanted, and I have no issues if they are what you want also. But when you get questioned as to why you want them, and the answer comes from your front cover, that is, who you think you should be, this is when you must re-evaluate your life choices. Now, I don't mean handing your goals to someone and they question you on them. Your goals are for you, you do not need to justify them to anyone. I mean when life questions you about your goals. If you're chasing the wrong things in life, life will give you plenty of warnings, and question you as much as possible, but if you are not aware of this, be prepared to live with the consequences. This may sound scary, but you need to know this. If your focus is in the wrong place, with the wrong intentions, that is, not aligned with your true self and your values, you must become aware of the path you are going down.

So, how do I set my personal goals?

Before we start, I want to make sure your environment is right to be able to do this. Your mind can easily become lost here, especially if you're overwhelmed. I recommend using the breathing technique here. The one you've learnt to use, to become more present. You may even need to find your peaceful place to do this exercise. Do whatever you need, to feel relaxed and to be free from distractions.

Let's take your values you have written down from earlier, also that list of what you imagine your life to look like. Remember I said we will need it later on? It will make sense now, as to why you kept it. Let's lay them out so you can see them. You will also need a variety of different coloured pens. With a blank sheet of paper, divide it into 4 even squares. Simply fold in half, and fold in half again. Each square will have the following headings; (1) 1-year, (2) 5-years, (3) 7-years, (4) 10-years.

At across the bottom of the page, I want you to write S.M.A.R.T. Then under each letter, write the following.

- Under S, write **Specific**
- Under M, write **Measurable**
- Under A, write **Attainable**
- Under R, write **Realistic**
- Under T, write **Timely**

This acronym will form the criteria in which each goal must meet before you write it down.

This part will require some thinking. Take the list of your life, next to each one of your potential memories, evaluate them, and ask, when would I like to achieve this by? Put a 1, 2, 3 or a 4 next to each them, to correspond with the time frames outlined in each heading. Keep in mind, that we can easily over estimate what we think is achievable in 12 months, but hugely under estimate what we can achieve in 10 or even 20 years. Start with your 4's. Ask the following question; what would need to happen for me to be able to achieve this? You're now looking for your 'vehicles' in which will take you to this goal. For farmers, generally speaking, your main vehicle will be your farm. So, what would the farm need to be doing by 10 years' time, to be able to achieve this goal. See how the responses to this question will come from a place of your values? Not your ego. Explore other vehicles too. You do not need to be restricted to your farm being the only tool to help you achieve your goals. Write down your answer in the 10-year box. Now, let's break it down. With the same colour pen, ask yourself, 'if that is what needs to happen in 10-years' time, what would need to happen by 7-years' time?' Write that down. In the same colour again, ask yourself, if that is what needs to happen in 7 years' time, what would need to happen by 5-years' time? Write that down. In the same colour again, ask yourself, if that is what needs to happen in 5-years' time, then what would I need to have achieved by 1 year from now? See how you've brought a 10-year goal, and broken it down to the next 12 months? The reason for using the same colour for this goal, is so you can visually link the steps together. I have seen this bring focus and success in so many farmers lives. Keep working on this exercise. Once you finish your 4's, choose a different coloured pen, move onto your 3's and keep breaking them down, then onto your 2's, break them down, then onto your 1's.

Let's say one of your values is Freedom. On your list of what you would like your life to look like, you have written down, a 6-month trip to Europe.

You decide that your 10-year goal is to have a 6-month holiday in Europe.

10-year goal: 6-month holiday in Europe.

Let's break that down now. What needs to happen in 7-years from now to achieve this?

Well, as I will be away in Europe for 6-months, you will need someone to be running the farm for me. Okay, so you'll need a farm manager hired, and they can learn your system for three years, and you're comfortable with them before you leave on your holiday.

So, the 7-year goal will be to hire a farm manager.

7-year goal: Hire farm manager.

Let's break that down again. So, if you want to hire a farm manager in 7-years' time, what will need to happen in 5-years' time?

I would want to make sure my farming model is profitable.

So, the 5-year goal would be to have a profitable farming model. For this goal to meet the criteria of S.M.A.R.T we will need to dive a little deeper here. What does a profitable farm look like to you?

I would like the farm to profit $500,000 on average.

See how the 5-year goal is S.M.A.R.T now? Let's write that down as our goal.

Remember, if an amount of profit is one of your goals, this is where the A in S.M.A.R.T comes into play. Make it attainable for you, without limiting yourself. Pushing your limitational thinking and achieving 90% of that, is so much more rewarding than setting low targets that you can reach without lifting a finger.

5-year goal: $500,000 profit.

Now, let's break this down. If this is what needs to happen in 5-years' time, what needs to happen in 3-years' time?

Well, to achieve a $500,000 profit I will need this much grain harvested per year. Considering the size of my land, I will have to average a yield of 4 tonne per Hectare of Wheat per year. We already average 3.5 tonne per Hectare of wheat so I would like to increase that by 500 kg per hectare in 3-years' time.

See how once you break goals down, they start to become realistic, and you start to think of ways to achieve them. So, the 3-year goal would be as followed.

3-year goal: Increase wheat yields by 500kg's per hectare

Lastly, break down the three-year goal to a bite size chunk that you can focus on for the next 12 months. If an increase of 500kg's per hectare of wheat is what needs to happen in 3-years'-time, then what needs to happen in the next 12-months?

I think I need to get an independent cropping advisor in to help create a plan for me so I have a systemised direction to follow. The 12-month goal will be as followed.

12-month goal: Hire an independent cropping advisor to help align my production system with my 3-year goal.

Look at what has happened here. We started with a 6-month trip to Europe and have brought it back to the next 12-months ahead. Using the farm as the vehicle to achieve this goal.

Once you finish this exercise, you should have a list of your 1-year, 5-year, 7-year and 10-year personal goals. Your focus now, should be on the next 12-months ahead. The most successful farmers I have seen do this exercise, only ever focus their attention on the next 12 months. They do not get caught up in how they will achieve the big 10-year goal, because they understand that if they achieve what they have set out to do over the next 12 months, they are one step closer to achieving their 10-year goal. Then, at the end of every year, they reassess their goals, make sure they are still aligned with their values and repeat the exercise. Setting out their next 12 months ahead. With your next 12-months set out for you, it is easy to jump from one goal to the next, and not completing any. You need to plan out specific steps you must take to achieve that specific goal. Then make time to work on them. this isn't hard, but it will take discipline on your behalf.

Take the 12-month goal from the previous example. In order to hire an independent cropping expert, you may consider the following steps:

1. Research local independent cropping advisors.
2. Contact a list of their clients for feedback
3. Shortlist three advisors
4. Meet with them one on one, interview style as they are potentially becoming part of your team, and get them to explain their processes they would use to achieve your goal.
5. Make an informed decision based on the data you have gained from the interviews
6. Work with the chosen cropping advisor to plan, budget and action the steps needed to achieve your goal.

See how there are six steps here. With twelve months to achieve them, it is easy to break them down into 2-months each. So, each step has on average eight weeks to be completed. The key here is to schedule it into your diary and be disciplined on its completion date.

The best way I have ever heard this explained is; Imagine you're in your car at night. You're going to drive from Melbourne - Australia, to Sydney – Australia. You put the destination into your GPS, it maps it out for you and you begin your journey. As it is dark, your headlights will only ever let you see 100 metres in front, at any given time. But you have full trust that the GPS will get you to Sydney. There is not one piece of doubt in your mind that you won't get there. This is exactly what goal setting is about. You've set the destination to where you would like to go. As you break it down, you have essentially mapped it out, just like the GPS, now all you have to do is focus on what the headlights will allow you to see, your next 12 months. Similar to that analogy, there should be no doubt in your mind, that if you keep focus on the next 12-months in front of you, you will eventually get there.

What happens if I set personal goals, and do not achieve them?

I love this question! I have been here plenty of times before! I have had some very ambitious 12-month personal goals previously. When my goals were coming from my front cover, falling short would get me very disheartened. It was a status thing for me. In 12-months' time I want to achieve this and this because then people will think I am successful. All my goals were ever doing, were feeding my ego. So, you could imagine what I would be like when I fell short. Set bigger ones and work even harder! That was my thinking. The snow ball effect would begin. I was always worried about HOW I would achieve the goal. But the how is none of our business, if you think 'how' when looking at your goals, you're starting to doubt their possibility. That is why I got you to break them down so all you can see is the next 12-months ahead. Trust the process and the HOW will come, you just have to have the awareness of it. This is where practicing your state of presence will have a huge effect. By practicing being present, when the opportunities come your way, you will be able to capitalise on them. Now my personal goals come from my true self, and they align with my values. I learnt that, what you value, will determine what you focus on. So, when I fall short of achieving a goal within the 12-month time frame, I am not hard on myself anymore. I simply ask why? Even though it is not fully complete, did it still move me towards my end goal? Is achieving this truly of value to me? I may have thought so 12-months ago, maybe I have changed and grown in the past 12-months? Will I commit to it for the next 12-months to see its completion? In my opinion, being ambitious and completing 90% of your value driven goals, will trump staying in your comfort zone and completing 100% of comfort zone goals. This is all for you to experience and play with. My advice is to just have fun with it, push the boundaries, feel what works and what doesn't for you. Maybe you need to start with only one goal a year and give that your 100% focus. Whatever you decide to do, I back you in all the way. There is a great book call 'The Gap and The Gain' by Dan Sullivan and Dr. Benjamin Hardy. It is a great read, and I think it suited for a lot of farmers. It explains that success of a goal should always be measure backwards. If you fall short of your goal, it is easy to be hard on yourself. This is because you're looking at it from where you thought you would be, not where you are at currently. If you fall short of your goal and look at everything you have achieved with that goal up to this point, you are focusing on your gains. So as in the book, I encourage you to always measure the success of your goal, backwards. A quote they use to explain this is, *"You can't connect the dots looking forward; you can only connect them looking backwards". – Steve Jobs.*

There is another reason why I would like you to set goals. A mentor of mine warned me on this, and it can be quite dangerous if you do it. It is the act of comparing yourself to others. I have been guilty of doing this before. I would look at the neighbours, or other farms in the district, and be disappointed that I wasn't up to their level. I didn't have the brand-new tractor, the amazing crops, or the productive pastures. In a way, I wanted to keep up with the Jones'. The issue here was, I would lose focus on what I actually wanted, or what actually was going to make me successful. By wanting these things, I thought they would make me happy. I also thought that they would make me look good and professional. At the time, I had a huge ego, my front cover was up, I didn't have any goals, no vision of what I wanted and no direction of where I was going. So, it was easy to look over the fence and say, I want that. Your happiness will never come from comparing what you have with what someone else has. Once you live following your values, what you truly want will become as clear as day.

Being inspired by others progress is the other side to this. I get a lot of my inspiration from what others have achieved. If you feel you're comparing yourself to others, change the lens and look to find inspiration from them. Be vulnerable and tell them you love what they have done. Ask them how they achieved it. Chances are they would be more than happy to share their experiences in accomplishing it. The most exciting part of all this, is that you do not need to reinvent the wheel here. Say you want to become fit. Start by finding the fittest person you know. Pick their brains, and find out how you can become as fit as them. This is all I have done with my goal to becoming a successful farmer. I didn't have to lock myself in a room and think about how I was going to achieve it. All I had to do was FIND and ASK! So, if there is something you would like to improve, or become an expert at, find people who have already achieved what you would like to achieve. Spend time with these people. If you have to pay them, then do it. I'm sure the results will pay for themselves ten-fold. The best question I have heard on this was a complete game changer for me. 'What is the one thing that you are not asking of someone, that if you did, could potentially change your life'. At the end of the day, the worst outcome is that they say no. So, you really have nothing to lose, but everything to gain by asking. Remember your goals will require action. You cannot create goals, and expect their completion through wishful dreaming. Once you set your goals, start to look for these people. They are your assets. They will give you great insight and direction in what it is you are wanting to achieve.

There are plenty of different ways to set goals, I am not saying this is the only way, but this is a start for you at least. If you find a way that suits you better, do it! Do not be afraid to experiment with this, in fact I encourage you to!

Pin your goals up in your office. Every time you achieve one of your goals, cross it out with a highlighter and celebrate! Give yourself a reward! I know one farmer who does this, and every time he achieves a goal with in the 12-month period, he gives himself a day off and goes fishing! This is a great way to keep motivated, especially when you feel tired or overwhelmed. Maybe you even need a coach to help stay motivated, like we talked about earlier. Goal setting may be completely new for you, you might not like the idea, and that's okay. I cannot force you to do anything. However, what do you have to lose by giving it a go? Who knows, it may just be the missing piece of the puzzle for you. I will leave you with this to think about. Your life is a journey, and *"planning a journey without a map, is like building a house without drawings."* – Mark Jenkins

The choice is yours.

Chapter Take-aways

1. 'If you chase two rabbits, you will not catch either one.'

2. Setting value driven goals will give you more focus to what you truly would like to achieve in life.

3. Use the goal setting exercise to breakdown, and get clarity on the steps needed in the next 12 months to see its achievement.

4. Review your goals yearly. Remember to measure their success backwards in what you have gained, rather than what you are yet to accomplish.

5. Focusing goals on a personal level first, will see the farm become the vehicle in which will be used to achieve them.

6. *'Planning a journey without a map, is like building a house without drawings'* – Mark Jenkins

Chapter 8

PRINCIPLE #7: LEARNING

We are all students

How would you rate your ability to learn? Give yourself a rating out of 10. 1 being poor and 10 being excellent. Be honest here. We all hold the ability to learn. But how well do you use that ability?

"A person who won't read has no advantage over one who cannot read." – Mark Twain.

Every day, no matter what you are doing, there are endless amounts of opportunities to learn. Do you see the opportunities? Or do you see the hassles? I was fortunate enough to understand this quite early in my career. The lesson was, if you are not moving forwards, you are going backwards. Along with, you cannot steer a parked car.

Let's take a look at the farming world 100 years ago. The early 1920's, and compare it to farming today. There is no way you would go back to using the equipment, the procedures or the techniques they used back then, and expect the same results that you are achieving today. As farming has evolved over time, technology has advanced and our results have grown. The same is with you as the farm-ER. The power of information you have access to today, there should be no doubt, that over the long-term average, your results should also too, be growing. I am talking about your personal results. Who you are today, is a better version of who you were yesterday. Sure, we have 'off days', like we have spoken about previously, but 'off days' do not ruin great work, they are merely just a moment in time. The biggest issues I have seen, over my journey, with the principle of learning, is the ability to 'unlearn'.

Learning is knowledge, and knowledge is power. Take this book for example. You could read this book from front to back, put it down, and keep on living your life the exact same way as usual. If someone were to ask you about it, you could say, 'it was very interesting, I learnt some valuable lessons on how to think differently, and techniques I can use in my day-to-day life'. That's all good and well to 'say' those things, but are you actually 'doing' those things. You definitely have more knowledge; I am not questioning that. However, I will question your ability to turn that knowledge into power.

Knowledge does not automatically turn itself into power. I could read one book after another. Listen to one podcast after another. Go to farming demonstration days, one after the other. Reality is, if you do not act on or implement anything you have learnt, then it stays a knowledge. I could easily talk about all the knowledge I learnt to others, but talk is cheap.

Anyone can do that. Anyone can repeat words, that's easy. Your power lies within your ability to implement.

I had a farmer pick me up one day and we went for a drive to a sheep production conference. This was early on in my career, so I did not know about the learning verses implementing theory. We got to the conference and he had a note book and pen. I felt unorganised because I turned up with nothing. We sat down and the presentation began. While I sat and listened, he sat and took notes, took photos of the presentation, and as I peeked over to have a look at his note book, a list of questions that he wanted answered. After the presentation had finished, I thought we would get up and head back home, maybe stop at the bakery or the pub, and make a day of it. NOPE! While everyone else was leaving, he got up and headed straight for the presenter. I could see him pointing to his note book, taking more notes, and deep in conversation with this presenter. On the way home, he asked me what I had learnt. With no notes, I had to rely on my memory, and you guessed it, all I could do was repeat the words of the presenter. I asked him the same question. What did you learn? He said, 'well, nothing yet, but I will let you know when I do.'

Wait, what? All those notes and he didn't learn anything? I was not satisfied with that response. I thought he would be talking the whole car ride home! I had to cure my curiosity here. I had to know why he didn't learn anything, yet used up two pens in notes. He went on to explain to me exactly what he had learnt from one of his mentors.

There are countless opportunities to learn, especially in your desired field in farming. If you really wanted to, you could go to a field day, conference or seminar once a month. Whether that be a farming topic or not. The options are endless. The trick is, to not become an addict of these things. I came here today because there were a few key issues I was having on my farm, and I chose this conference because I believed that I could get answers. Turns out I did get answers. Are they the right answers? I am not sure yet, that will be answered later on in the results. What I do know, is that if I came here looking for answers, and did not implement them when I got home, then I would have wasted my time. My time is too valuable to go to conferences or seminars for the sake of just going. I go with the intention to take one key thing away from them, and implement it. It may not always work out, but there is also a similar chance that it will improve my business. This is why you do not want to become addicted to them. If you're thinking, 'I will implement one thing from every conference or seminar', you will overload yourself, and be focusing on too many things at once.

He put it simply. I look at what I would like to achieve for the year. Then research conferences, seminars and books that I believe would be an asset in helping me achieve those outcomes. Likewise, if I have an undesirable outcome on the farm, and want to fix it, I will research the same again, and find one that will help point me in the right direction to improve. I only ever commit to one conference or seminar per quarter, one every three months, which will give me four for the year. That is my limit.

I liked this; I was getting a real insight into the tools that this successful farmer was using. He was leveraging off other people's information, to improve himself.

The 48-hour rule:

There were two rules he taught me around how to learn. The first one was the 48-hour rule. It was quite simple. You have 48 hours to start to implement your new learning. Have you ever been told some thrilling news or done something extortionary, like your offer has been accepted on that new farm, or you went skydiving? In that moment you are full of energy, your body thrives with excitement, but somehow 2 days later, it isn't that exciting, it almost becomes old news, and things are back to the way they originally were. Sound familiar? Well, this is the same when you learn something new. Your mind is racing about, thinking of all the different possibilities that this new learning, could potentially be. You're excited, full of energy, but 2 days later you're back into your old ways, your old habits and nothing has changed. The 48-hour rule is about capturing that enthusiasm. Invest in that enthusiasm by beginning to take action with your new learning. By taking action when you are on a high, you allow it to compound, and this is when you start to see your new learning come to life.

Say you go to a feed lotting conference because you're looking to upscale your production. You get a load of ideas on how this could increase your business, and you're excited about the potential possibilities it could bring. When you arrive back home you are tired and ready to relax. Waking up the next morning, you're still full of excitement but there are catch up jobs needed doing as you had yesterday off. You get caught up on the farm jobs, that the feed lot idea is pushed to the side. Before you know it, a month has passed by and you're no closer to achieving any of the outcomes you desire. In fact, it was probably a waste of a day going to the conference. You lost all momentum. As the 48-hour rule states, you need to keep building on the momentum from the conference. Instead, you come home, tired, but decide to start brain storming some ideas. Possible locations, a list of people who are doing this that you could talk to, the next steps to take and by when you want to take them by. This all helps to keep momentum. You may get three months in and find that this idea isn't worth perusing, however, the knowledge you gained in the process, helped determine your decision making. It wasn't that you convinced yourself it was too hard through taking no action. It was a thought-out process that you done your due diligence on, that just so happened to not fit into your farming system.

The second rule was unlearning. How often do you see someone, doing a task a certain way, but having no idea as to why they do it that way. This fascinates me. Especially in farming when I see it play out. There is no wrong or right here, this is just human nature. This is the lens that they are seeing the world through. There is a story about this, called 'Grandma's Ham'. I will paraphrase here; however, a man watches his wife prepare a ham for dinner. He watches her cut off one inch from the ham at either end. He asks, 'why did you cut that off?' she replied, 'that is how mum use to do it'. later that day, the ladies mother arrives for dinner, and he decided to ask her, 'why do you cut one inch off each end of the ham?' She replied, 'Well, that is how my mother use to do it'. Curious to know why, he called his wife's grandmother and asked her the same question. Her response, 'I did that so it would fit into my baking pan'. See what can happen when we do not question our methods? The world around us moves forward, whereas we stand still.

The farmer told me that in order for him to implement something new, he must first unlearn his old ways. The easiest way to unlearn your old ways is to first understand them. Understand why you do something the way that you do. Understand why that has become a habit for you. Where did it come from? Where did you first see it?

To swap out an old way with a new way takes an open mind. It may not be wise to just jump in the deep end here, then decide to learn how to swim. That may be how you operate, and that's okay, go for it! But, if you need to see results to be able to unlearn first, then dip your toe in, see the results of the new way, once you see the results, and are happy with them, then you can go fully in.

How would you rate your ability to learn now? Give yourself a rating out of 10. 1 being poor and 10 being excellent. Be honest here. We all hold the ability to learn. Considering what you have just read, are you putting your ability to full use?

I would like to focus on learning at a personal level now. For you, the farmer. If you're unhappy with the way things are at the moment, you have two choices. Either you make a change, or you put up with it. The most successful farmers I have met, don't put up with anything that does not bring them joy. They are open to change. They embrace change. They love change. They make the time to learn new things. They are excited to learn and are open to sharing their results. When leaning comes from their values, it becomes inspiring to them, not a chore. You can chase farming outcomes for years on end. But what are you learning about yourself? For yourself.

This idea of personal learning is very new to me. I thought reading self-help books or listening to podcasts was enough. But I had it slightly wrong. I would only read books or listen to podcasts if they were going to teach me how to make money, remember, I once defined my success with money. I soon discovered that the personal learning journey is so much more. Discovering who you really are and what brings you joy and happiness. Releasing the pressure off having to achieve and to have a little more fun along the journey. I always said to myself, 'when I achieve this, then I'll be able to do that.' What if achieving that one particular thing was going to take me ten years. I would sacrifice everything to achieve it, especially my happiness. What's the alternative I see now? Well, now I still want to achieve plenty! But I make sure I enjoy life whilst achieving it. Trust me, a weekend at the beach won't stop you from achieving your goals. In fact, it will more than likely fast track it!

I even make sure I have multiple 'laughs' during the day. Especially at myself when I make a mistake. I know for a fact that getting angry at myself won't help, so I may as well find the funny side in the situation.

My goal is that this book has started your self-learning journey, or at least giving you a positive direction to move towards. It is easy to stick to your old ways, stick to your old habits, believe that your way is the only way, or that you don't actually need to learn anything new. Let me leave you with this famous saying to think about; *'for things to change, first I must change'*.

Chapter Take-aways

1. Learning is knowledge and knowledge is power, but only if you take action on it.

2. There is a trap of information overload when you become and addict to conferences and seminars. Choose the ones that will help either you or your business move forward.

3. Aligning your 12-month goals to conferences or seminars is a great way to keep motivated and help with their achievement.

4. Use the 48-hour rule to keep momentum growing when you learning something new and of value to you.

5. You do not need to sacrifice your happiness in order to achieve a goal. A weekend at the beach will not stop you from achieving success. In fact, it will most likely fast track it.

6. *'For things to change, first I must change.'*

Chapter 9

LET'S TALK LEGACY

How would you like to be remembered?

This isn't a key principle; however, it was something that was very common with all successful farmers I had spoken to. It is the legacy they were creating. What they wanted to leave behind and how they wanted to be remembered. Everyone's legacy was unique to them. From internal family legacy, to a global leading product legacy. It was an interesting concept that I hadn't heard much about before, but it had my full attention. I wanted to know why all these successful farmers wanted to leave a legacy, and why, in my eyes, some were huge and some weren't.

I remember meeting up with a farmer one day. I had reached out to him because he has a very successful farming business and a very successful off farm business. I had the intention of asking about his legacy, where it comes from, and what it means to him. Deep in conversation, I was blown away by how much his legacy lit him up inside. I could see that this was bigger than just him. It came from pure selflessness. He went on to say that it is his 'WHY'. It is the reason why he gets out of bed every morning; it is the reason why he puts in the effort; it is the reason why he never gives up.

So, with my curiosity skyrocketing, because I thought you just got out of bed in the morning anyway, I had to find out more. I found a book call 'Start with Why' by Simon Sinek. I recommend reading it if you get a chance. He explains the whole concept of 'why' and its importance in your personal life, but also your business. To summarise my key take away from the book, it is all about purpose. Your purpose. Understanding what you do is great, but understanding why you do it is amazing! If you do not know why you do something, you essentially live other people's lives. Other people will just use you for their purpose. Now your purpose in life isn't to just get to the other end. That's boring. That's not what life is meant for. Your purpose is much more than that. You are much more than that! I'm not saying you have to be the next Bill Gates and Warren Buffet. What I am saying though is you need something. A reason. A reason that is bigger than you. A purpose that helps others. Let's look at two farmers here, both producing the same product. One with a purpose, and another who hasn't found their purpose yet. The farmer with a strong purpose, wants to create a product of the highest quality, that can be used globally, that is enjoyed by consumers and is ethical and sustainable. This farmer can see the big picture, as clear as day. It excites them, it is the core focus when decision making, it aligns with their values, they could talk passionately about it for hours on end, it puts a smile on their face just thinking about it. How do you think this farmer is in their day-to-day life? Focused? Relaxed? Happy? I would say so. It's not as if they don't have bad days. It's just they have a bigger purpose, so when a bad day comes around, they can use their purpose to stay focused, and not let it become a distraction. Let's look at Farmer #2 now, who hasn't found their purpose yet. Now remember they are producing the same product here. This farmer still wants to produce a product of high quality, not for the consumers sake, but for their pocket. Money is their objective, and what happens when money becomes the objective? Everything else goes out the window. They do not care where it ends up, who their consumers are, and the words 'ethical' and 'sustainable' can be replaced with 'whatever makes me the most money wins'. Now, I am all for making a profit in farming, I think we should all be aiming for that. However, purpose driven operations, with purpose driven operators win hands down every time, over the long term, versus the cash grabbing model that is driven with no purpose. The farmer #2 could also have a 'don't care' attitude. They struggle to get out of bed in the morning. They fall behind, or are late with executing farm tasks. It is not their fault; it is simply that they do not have a purpose. I believe once you find your purpose, your life will become full of happiness. So, what is your purpose? Why are you doing what you're doing?

I struggled with this concept for years. When I first started learning about this, I thought I had to know my purpose straight away. If I didn't have a purpose, then I wouldn't be successful, and that is the last thing I wanted, to not be a success. I would have a purpose for a week or two, then watch an inspiring YouTube video on something, and go, 'I like that, actually I change my purpose to that now'. I was in this spiral of claiming my purpose, based on everything and everyone around me. It was coming from a place of want and need. I was in the trap of comparing myself to others, mind you I was comparing myself to people who had 20 plus years' experience on me. This is not a great place to be. Looking back now, not one of those made-up purposes gave me any long-term excitement, or happiness when thinking about what they could become. I think I was picking them, and jumping between them, all because I was worried about what would happen if I did not have a purpose in life. This did eventually change for me though. I was given a copy of another Simon Sinek book, called 'Find your Why'. Again, I highly recommend reading this if you are struggling with finding your purpose in life. As I was reading through the book and doing the exercises, I was having light bulb moments, going off in my mind. I started to see things that I hadn't seen before. I started to link childhood memories to what motivates me today. It was an amazing insight to finally understand the concept of a purpose. It has helped shape my decisions and drives me to get out of bed each morning. My purpose, is to help farmers find success and succeed in their lives.

If you know your purpose in life already, and know what type of legacy you'd like to leave, then that's amazing! However, if you are struggling to find your purpose, that's okay. Again, I recommend reading 'Find your Why' by Simon Sinek. Chances are, you are already living your purpose. You just haven't taken the time to define it yet. If you're living each day, through passion for what you do, excitement for the future in what you're doing, and joy to do it all again the next day, then, in my opinion, you have it! That is your purpose right there. The legacy you create from that purpose is up to you. Your legacy could potentially help one person, or one million people. There is no wrong or right answer. Whatever you choose is correct. I like to think of a legacy this way. What is missing today, that would enhance the life of someone else's tomorrow?

Chapter Take-aways

1. What is the legacy you are creating?

2. Use the book, 'Find your Why' by Simon Sinek to define your why. The reason you get out of bed every morning.

3. Purpose driven farmers are unstoppable.

4. You should be living each day through your passion which brings you joy and happiness.

5. Your legacy is special to you, own it.

6. Ask yourself, what is missing today, that would enhance the life of someone else's tomorrow?

Chapter 10

ROUTINES FOR SUCCESS

You are the results of your habits

This again is not a key principle, but a theme I found too common not to leave it out of this book. It is the practice of a morning and night routine. When I was learning about routines, I was recommended a book called, 'The 5am Club' by Robin Sharma. His theory with a morning routine is Exercise, meditate, learn. 20 minutes on each, giving you an hour-long morning routine. I gave this a go and really enjoyed it. By 6 am in the morning, I had completed a 'power hour' focusing on only myself. My biggest lesson from performing this was self-discipline. A characteristic I believe all successful people hold. Especially on the cold and wet mornings, when you can hear the rain, however, you get up anyway, embrace the uncomforting cold, and get on with it. We ultimately have to pains we can live with. That is the pain of discipline, or the pain of regret. To implement a morning and night routine, I truly believe you first need to understand why you are doing it. What are the benefits of this to me? Once you understand why, then implementing becomes ten times easier. There has been plenty of research done on the benefits of morning and night routines, but the one I like the most, is 'setting the tone'. By waking up an hour earlier, and investing time into yourself, you are setting the tone for your day. This is all about control. You are in control when you make the time for you. When you give yourself an hour to get ready for the day ahead, you are in no rush, you could easily still be in bed sleeping, but you have chosen to practice self-discipline instead.

Let's look at introducing a morning routine into your life. I think it is important here to take baby steps when you are beginning. Say you wake up at 6:30 am every morning, set your alarm for 6:20 am to start with. Give yourself an extra 10-minutes in the morning. That extra 10-minutes is for you to get up, go to your values on your bathroom mirror or background wallpaper on your phone, and read them out to yourself, and feel them coming to fruition in your body.

Do this until waking up at 6:20 am and performing this task becomes second nature. The time this process takes is up to you. Research says that forming a habit can take on average up to 66-days. Which isn't really that long when you consider 66-days of discipline for thousands of days of benefits. Once you get comfortable with this new habit, add something new in. Go for a 6 am wakeup now, perform your natural 10-minutes of values routine, then 20 minutes of reading for example. Keep this up until it becomes part of your morning, then add something else into the mix. You may wake up at 5:30 am now, perform your 10-minute values routine, your 20-minute reading routine, and then a 30-minute walk. All of a sudden you have a morning routine that has you waking up an hour earlier, filling it with tasks that bring you joy, and setting the tone for your day to be quite positive! I think it is quite powerful to be in control of your day. I recommend to have fun with this! The whole purpose of this is to bring you joy and happiness. Be okay with taking something out to add something new in. Be okay with changing it daily, for example if it is pouring rain, you may decide to do an indoor workout rather than going for a walk. It is important to find what works for you, and to find your limit. If you commit to 1-hour each morning, fill that hour with 2-3 tasks that you can give your whole focus and attention to. Less is more in this space. If you feel you need to extend your hour, then give it a go. If you're enjoying it then keep it up. However, if it does not feel like it is working, maybe it is taking away from other areas of your life, then you may need to consider dropping back down to an hour.

You will need to experiment with your wake-up time also. If you're waking up at 6 am every morning, feeling fresh and powerful, then decide to change it to a 5 am wake up, feeling tired and unmotivated, I would say keep it at 6 am. Rearrange your morning to fit in your routine with your normal wake up time. However, this may be bed time issue as to why you're so tired and unmotivated waking up earlier. I would also consider adding in a 'day off'. Every Sunday morning you may decide to have a sleep in, no alarm, just wake up when you wake up, and your routine may be as simple as a cup of tea outside enjoying the sunshine or listening to the rain on the roof. I am saying all this because a morning routine is something to look forward to, something to enjoy and something you wouldn't want to miss out on.

I spoke to a farmer and a mentor of mine one day, and he was telling me about his morning routine. He said, as soon as he wakes up in the morning, he asks himself one simple question. 'How am I feeling?' This then allows him to decide what his morning will look like. He called it his 'tool box of morning routine tasks'. If he is feeling energetic and motivated, he will do a physical workout. If he is feeling full in his mind, he will grab his journal and start writing everything down. I liked this concept. He does not need to commit to any specific task, all he commits to is his wake-up time, and seeing what his body needs. Genius if you ask me!

There is a flipside to what I have just explained, that is self-discipline. If you are wanting to create the habit of self-discipline through your morning routine, then it will look a little bit different to what you have just read. I do not intend to scare you, but this is not for everyone, and will not fit into everyone's lives depending on where you are at. I remember when I was creating this self-discipline habit, the most common comment I got was, 'it's clear you don't have kids.' I would get a good laugh out of it, but I am grad I went through that period of practicing self-discipline.

When setting up your routine, you are to pick a wake-up time and lock it in. If you choose 5 am, then 5 am it is,

7-days a week. No snooze button, no excuses. I remember my alarm going off some mornings and thinking, 'surely not already!'

All I would do was the 5-second rule. I knew the right decision was to get up out of my comfy bed and embrace the uncomfortable. I would put a smile on my face and start counting. 5, 4, 3, 2, OUT. I always made sure I was out of bed before I got to 1. I am not sure why I done this, maybe the fear of regret of getting to 1 and staying in bed for an extra hour.

You must also have your routine set out perfectly. It should be like clockwork. For example, you may have three tasks in 1 hour to do. So, you allocate 15 minutes to each task, allowing 5 minutes before each task to get ready. This is all about creating and understanding self-discipline. This doesn't have to be something you do forever, but if you're even a tiny bit intrigued, I would recommend giving it a go for sure!

Now onto night routines. These are well and truly more relaxed. This is a time for winding down, reflection, and setting yourself up for tomorrow. This could be as simple as the 30-minutes before you hop in bed. Say your bed time is 10 pm. So, at 9:30 pm, you decide to put your phone away and journal about your day for 5-to-10-minutes. Reflect on everything that went well, everything that could have gone better, and what your learnt. This is important to let these things go as they are history now; tomorrow is a new day. You prepare anything you may need for your morning routine tomorrow, brush your teeth, and hop into bed. Once in bed, we want to make sure we end on a high and positive feeling. The best way I have seen this done it to say three things you're grateful for today, and say three things you're looking forward to about tomorrow. This is always great to leave you feeling happy and excited, then you fall asleep on that.

Have fun with implementing these into your life! Enjoy the process of finding out what works, and what doesn't work for you. All I have known to understand, is that successful farmers all take the time to set the tone of their day before it starts, and to reflect on their day before it ends.

Chapter Take-aways

1. Implement a powerful morning routine that will set the tone for your day ahead.

2. Have a variety of different tools you can use each morning to help perform your routine depending on how you are feeling.

3. A self-discipline morning routine is not for everyone, but is a great way to build discipline into your life.

4. Implement a night routine to help relax you and finish your day on a positive note.

5. End your day with three things you're grateful for, and three things you're looking forward to the most for tomorrow.

6. Have fun and experiment with this!

Chapter 11

BRINGING IT ALL TOGETHER

"Action is the foundational key to all success"
– Pablo Picasso

Over my journey of conversations with successful farmers, I discovered that success isn't something that they necessarily pursued. It just so happened to be a result through their daily habits. They attract success. By practising the seven key principles, in their own individual way, they improved numerous areas within their lives. Key areas such as career, health, family, friends, growth, finances and fun. By finding happiness and balance in these areas, success was an inevitable part of their results. They all understood one thing. The most value piece of real estate that they own, isn't their farm, but is the six inches of space between their ears. Their mind.

This book can be a lot to take in all at once. It can be quite easy to become overwhelmed with all this information. When we have an information overload, we tend to shut down and ignore it, because that is what feels most comfortable to us. My challenge to you is, find one principle or key takeaway from this book that really gave you some inspiration to change. Focus on that one for now. Form a new habit around that one particular principle. Once you're comfortable with it, and it is now a part of your daily routine, add another principle into your life, then another, and another. Keep repeating this process until all seven principles are effectively working in your life. This will allow you to only focus on one thing at a time. We need to remember to crawl before we walk. This journey is a marathon, not a sprint. This will be something that you must work on daily. Sorry, but there are no short cuts here. You have to put the effort in and do the work. No one can do it for you. The question I have for you is, 'how happy are you with who you are, and the results in your life today?' If the answer is you aren't satisfied with who you are, and aren't happy with your results, then what are you going to do about it? There will come a breaking point when the pain of staying the same, outweighs the pain of change. This may not have happened to you yet; however, you may feel this is the path you're on at the moment. My advice to you is, be proactive, start to make those small changes in your life today, start to create the new version of you. These principles will take on new meaning as you grow, and develop through the different stages of your life. So be okay with changing and rereading this book to adapt to your new way of thinking. The best thing you can do is to make a start. I guarantee you the results will come your way; it all starts with you.

ACKNOWLEGMENTS

I would sincerely like to thank you for taking the time to read this book. I am truly passionate about the success of the farmer, and have enjoyed being able to help you along your journey towards success. My aim is for you to start to think differently, and I hope this has made you want to make changes in your life. If it has motivated you to make a change, feel free to recommend this book to someone you know who could use it. We all hold the ability to make positive changes in our lives. The ones you have read about in this book are free. They will cost you no money to implement them into your life. Some people start doing this, and do not see any results in the first two months, so they give up. I encourage you to keep pushing, keep practising daily, and the results will inevitably come. Trust and enjoy the process. This is your journey. This is your story you're writing through farming and life. Own it. I will leave you with this.

'Don't fall in love with success, fall in love with the habits that bring success.' – Unknown.

RESOURCES

- The 7 Habits of highly effective people - Stephen R. Covey
- The Secret - Rhonda Byrne
- 5 Second Rule – Mel Robbins
- From People Pleaser to Soul Pleaser – Tracy Secombe
- Mark Kluwer – Breath and Breakthrough
- Kane Johnson – Breath and Breakthrough
- Trevor Hendy – Breath and Breakthrough
- Happiest Man on Earth – Eddie Jaku
- Slight Edge – Jeff Olsen
- Andrew Roberts – Co-Founder of Farm Owners Academy
- 'Start with Why' & 'Find your Why' – Simon Sinek
- 5am Club – Robin Sharma
- The Gap and the Gain – Dan Sullivan & Dr. Benjamin Hardy
- Farm Owners Academy – farmownersacademy.com

I would like to thank everyone who has been a part of my journey so far. To all the farmers, my partner, parents, mentors and coaches. You have all played a huge role in helping me to bring this book into fruition. Thank you for your ongoing support. To the reader, thank you for the opportunity for me to share my experiences with you. I am grateful I get to help elevate the lives of farmers all around the world.

www.ingramcontent.com/pod-product-compliance
Lightning Source LLC
Chambersburg PA
CBHW070157100426
42743CB00013B/2953